# Financial Management of Flood Risk



OECD
BETTER POLICIES FOR BETTER LIVES

**Please cite this publication as:**
OECD (2016), *Financial Management of Flood Risk*, OECD Publishing, Paris.
*http://dx.doi.org/10.1787/9789264257689-en*

ISBN 978-92-64-25767-2 (print)
ISBN 978-92-64-25768-9 (PDF)

Co-published by IWA Publishing
Alliance House, 12 Caxton Street, London SW1H 0QS, UK
Tel: +44 (0)20 7654 5500, Fax: +44 (0)20 7654 5555
publications@iwap.co.uk
www.iwapublishing.com

ISBN: 9781780408569 (Paperback)
ISBN: 9781780408576 (eBook)



**Photo credits:** Cover © apirati333 / Shutterstock.com

Corrigenda to OECD publications may be found on line at: *www.oecd.org/about/publishing/corrigenda.htm*.

# *Foreword*

*Disasters present a broad range of human, social, financial, economic and environmental impacts, with potentially long-lasting, multi-generational effects. The financial management of these impacts is a key challenge for individuals and governments in developed and developing countries. G20 Finance Ministers and Central Bank Governors and APEC Finance Ministers have recognised the importance and priority of disaster risk management strategies and, in particular, disaster risk assessment and risk financing.*

*The OECD has supported the development of strategies for the financial management of natural and man-made disaster risks, under the guidance of the OECD High-Level Advisory Board on Financial Management of Large-scale Catastrophes and the OECD Insurance and Private Pensions Committee. This work has included the elaboration of an OECD Recommendation on Good Practices for Mitigating and Financing Catastrophic Risks and a draft Recommendation on Disaster Risk Financing Strategies to update the OECD's guidance in this area, as well as a number of global events aimed at sharing experience on approaches to disaster risk financing and identifying key challenges where international cooperation would be beneficial. In cooperation with other international organisations, the OECD has also responded to requests from the G20 and APEC through the development of the* Disaster Risk Assessment and Risk Financing: A G20/OECD Methodological Framework *and a report on* Disaster Risk Financing in APEC Economies: Practices and Challenges. *In 2015, the OECD published* Disaster Risk Financing: A Global Survey of Practices and Challenges *which provides an overview of the disaster risk assessment and financing practices of a broad range of economies relative to the guidance elaborated in the* Disaster Risk Assessment and Risk Financing: A G20/OECD Methodological Framework.

Financial Management of Flood Risk *extends this work by applying the lessons from the OECD's analysis of disaster risk financing practices and the development of its guidance to the specific case of floods. This report was prepared by the OECD Secretariat based on input provided in response to an OECD survey questionnaire as well as research undertaken by the OECD and other international organisations. The report provides an overview of the approaches that economies facing various levels of flood risk and economic development have taken to managing the financial impacts of floods. The report benefited from the support and input of the OECD High-Level Advisory Board on the Financial Management of Large-Scale Catastrophes and the OECD Insurance and Private Pensions Committee.*

# *Table of contents*

## Tables



## Figures



## Boxes

# Abbreviations and acronyms

| | |
|---|---|
| **CCR** | *Caisse centrale de réassurance* (public reinsurer in France) |
| **CCS** | *Consorcio de Compensación de Seguros* (public insurer in Spain) |
| **CCRA** | Climate Change Risk Assessment (United Kingdom) |
| **CRED** | Centre for Research on the Epidemiology of Disasters |
| **CRS** | Community Rating System (United States) |
| **EAD** | Expected Annual Damage |
| **EM-DAT** | Centre for Research on the Epidemiology of Disasters' Emergency Events Database |
| **FEMA** | Federal Emergency Management Agency (United States) |
| **FIRM** | Flood Insurance Rate Map (United States) |
| **GAO** | Government Accountability Office (United States) |
| **GDP** | Gross Domestic Product |
| **ICI** | Iceland Catastrophe Insurance |
| **IPCC** | Inter-Governmental Panel on Climate Change |
| **NFIP** | National Flood Insurance Program (United States) |
| **OECD** | Organisation for Economic Co-operation and Development |
| **SFHA** | Special Flood Hazard Area (United States) |
| **WTS** | *Wet Tegmoetkoming Schade bij Rampen en Zware Ongevallen* (disaster compensation law in the Netherlands) |

# Executive summary

Flooding is one of the most common, widespread and destructive natural perils, affecting approximately 250 million people and causing USD 40 billion in losses on an annual basis. The increasing accumulation of assets in floodplains and coastal zones, combined with the expected impacts of climate change on precipitation patterns and sea levels, are likely to lead to increasing losses in the future. As a result, significant policy attention is being focussed on finding ways to effectively manage the financial impacts of flood risk, considering the roles of investments in risk reduction as well as mechanisms for transferring flood risk. Insurance and other risk transfer tools can make an important contribution to the financial management of flood risk by spreading the risk across domestic and international (re)insurance and capital markets and reducing the share of losses absorbed by households, businesses and governments.

There is a wide variety of approaches across countries to protecting households and businesses against the financial impacts of floods, designed with the aim of achieving different policy objectives, such as broad availability and affordability of coverage, solidarity in terms of loss-sharing across regions, the establishment of clear incentives for risk reduction and/or significant transfer of risk to private markets – with clear trade-offs between these different objectives. In some countries, flood insurance arrangements have led to broad coverage of flood damage and losses although this is far from universal. Overall, a significant financial protection gap remains which leaves households and businesses – and ultimately governments – exposed to substantial risk of financial losses.

There are a number of important impediments to the insurability of flood risk in many countries, including the size of potential losses from a flood event, the ability to establish a diverse pool of insured risks as well as the level of uncertainty in estimating potential losses, particularly in the context of a changing climate. A number of countries have established innovative approaches to addressing these challenges by investing in risk reduction at the community and household level, improving the quality and availability of flood risk maps, and enhancing public awareness of the risk of flooding and the need for financial protection. To complement these direct investments, many countries are also examining ways in which communities and households can be encouraged to protect themselves against flood risk, including by ensuring that public sector risk reduction investments and insurance and financial assistance arrangements do not discourage private initiative.

Governments in flood-prone countries face significant costs related to the financial management of flood risk, including the costs of investing in *ex ante* risk reduction as well as *ex post* costs related to emergency response, reconstruction of public assets, and compensation and financial assistance to sub-national governments, businesses and individuals affected by floods. Where insurance coverage is more limited – whether due to specific challenges in providing coverage for flood risk or broader challenges related to the level of insurance market development – governments will also face significant pressure to provide compensation to those affected. A careful assessment of the relative costs and benefits of different approaches to managing these costs is critical for the effective financial management of flood risk.

## Surveyed countries

The report benefitted from responses to a questionnaire from 27 countries from across the world, facing very different levels of flood risk and insurance market development: Australia, Austria, Canada, Chile, Costa Rica, Czech Republic, Estonia, France, Hungary, Iceland, Ireland, Israel, Japan, Latvia, Mexico, Myanmar, New Zealand, Peru, Philippines, Poland, Portugal, Russia, Spain, Switzerland, Turkey, United States and Viet Nam.

## Key findings

- *The ability to quantify exposure to flood risk, including in the context of a changing climate, is a prerequisite to the effective financial management of flood risk and a necessary input for assessing the costs and benefits of different approaches to risk reduction and for transferring risk to (re)insurance and capital markets.*
  - However, a number of challenges remain in terms of the quality and consistency of flood risk maps and the coverage of probabilistic flood models. In many countries, the development of private flood insurance markets has been a key driver for the development of flood modelling capacity. Information on past events, including from the insurance sector and satellite imagery, can provide an (imperfect) alternative source of information on potential exposure where probabilistic flood models do not exist.
- *There are a number of important challenges to the insurability of flood risk which have led to a significant "financial protection gap" in terms of the insurance coverage of flood losses and damages.*
  - Government involvement is key to supporting the insurability of flood risk through effective land-use planning and investments (or encouraging investments) in risk reduction at the community and household level. A number of countries specifically link issues of insurance availability and affordability to decisions on land use and flood protection investments.
- *The form of insurance coverage can have important implications for take-up rates and the incentives created for risk reduction.*
  - Insurance arrangements that make it more difficult for policyholders to exclude flood coverage in their general property insurance policies have been more successful in achieving higher levels of flood insurance penetration. However, whatever the form of insurance coverage, the contribution of insurance to the financial management of flood risk will be maximised where insurance promotes risk awareness and risk reduction.
- *Effective coordination across government is critical for establishing an integrated approach to the financial management of flood risk that considers the best use of public resources, and takes into account the costs and benefits of different approaches (including the incentives created by different interventions).*
  - Given the range of policy tools that need to be considered, a holistic approach to the financial management of flood risk requires effective coordination across government, including across levels of government, supported by strong leadership aimed at addressing the financial vulnerabilities created by exposure to flood risk.

*Chapter 1*

# Introduction: The prevalence of flood risk

*This chapter provides an introduction to the evolving nature of flood risk and the implications for the financial management of that risk. It also provides an overview of the structure of this report.*

Flooding is one of the most common, wide-reaching and destructive natural perils, affecting, on average, approximately 250 million people around the world each year (UNISDR, 2013). Practical considerations such as access to water supplies, fertile soil, waterborne transport and the attractiveness of living near rivers and coasts have historically led to significant development in areas prone to flood risk. In many countries, substantial portions of the population now live in areas prone to flooding; for example 49% of the population of Japan is located in former river and coastal floodplains (Sato, 2006) and two-thirds of the population of the Netherlands lives in flood-prone areas (Jones-Bos, 2011). A number of mega-cities in Asia, including Ho Chi Minh City, Jakarta and Manila have been repeatedly affected by flooding in recent years. In the United States, floods accounted for almost two-thirds of all presidential disaster declarations during the period 1953–2010 and have been responsible for the largest number of lives lost and the most damage over the last century when compared with other natural disasters (Michel-Kerjan and Kunreuther, 2011). In Canada, floods have accounted for 40% of all recorded natural disaster events since 1900 (Insurance Bureau of Canada, 2015).

Population growth and the accumulation of assets in flood-prone areas have led to a substantial increase in built-up areas susceptible to flooding and therefore the size of the impacts arising from flood disasters. According to some projections, more than half of the world's population is expected to live within 100 kilometres of the coast by around 2030 (RMS and Lloyd's, 2008). The frequency of flood disasters is likely to increase as the number of people exposed to floods is expected to grow at a higher rate than general population growth (Keating et al, 2014). Increasing urbanisation will exacerbate this trend as, in urban areas, the capacity for rainfall absorption deteriorates and water runoff increases significantly above what would be expected to occur on natural terrain.

While subject to significant uncertainty, climate change is also expected to have an impact on the level of flood risk through changes to precipitation patterns (such as a higher incidence of heavy precipitation events), increases in coastal inundation as a result of sea-level rise and changes to the range and intensity of tropical cyclones and hurricanes. When taking into account the potential impacts of climate change, an estimated 147 to 216 million people could live on land that is below sea level or below regular flooding levels by the end of this century (Climate Central, 2014).

Annual average losses from flood events have increased to an average of over USD 40 billion annually in recent years. While a significant component of the increase in losses relates to increasing asset values and better reporting, there is some evidence that the frequency of flood disasters has also increased. The number of reported flood disasters throughout the world nearly doubled in 2000-2009 relative to the previous decade and more flood disasters occurred between 2010 and 2013 than in the whole decade of the 1980s (Keating et al, 2014).

More than 75% of the countries that responded to an OECD survey questionnaire perceive themselves as facing moderate or high levels of flood risk from inland flooding (including over 30% that perceive themselves at high risk). In the case of coastal countries, just under 50% indicated that they face moderate or high levels of flood risk from coastal flooding (see Table 1.1).

Table 1.1. **Perceptions of flood risk**

| Country | Inland Floods | Coastal Floods |
|---|---|---|
| Austria | Moderate to High | N.A. |
| Canada | High | Moderate |
| Chile | Moderate | High |
| Costa Rica | Moderate | Low |
| Czech Republic | High | N.A. |
| Estonia | Low | Low |
| France | High | High |
| Hungary | Moderate to High | N.A. |
| Iceland | Low | Low |
| Ireland | Low | Low |
| Israel | Low | No risk |
| Latvia | Low | Low |
| Mexico | Moderate | Moderate |
| Myanmar | Moderate | Moderate |
| New Zealand | Moderate | Low |
| Peru | High | Low |
| Philippines | High | High |
| Poland | Moderate | Low |
| Portugal | High | Moderate |
| Russia | Moderate | Moderate |
| Spain | Low | Low |
| Switzerland | Moderate | N.A. |
| Turkey | Moderate | No risk |
| United States | Moderate | High |
| Viet Nam | High | High |

*Note:* For the purposes of the survey, a high level of risk indicated that a significant proportion of the population (more than 10%) is vulnerable to frequent flooding (with an expected return period of 1 in 75 years or more frequent); a moderate level of risk indicated that a significant proportion of the population (more than 10%) is vulnerable to occasional flooding (with an expected return period of between 1 in 75 years and 1 in 200 years); and a low level of risk indicated that only a small proportion of the population (less than 10%) is vulnerable to infrequent flooding (with an expected return period 1 in 200 years or less frequent). "N.A." was assigned to land-locked countries that face no risk of coastal flooding.

*Source:* Country responses to an OECD questionnaire on the financial management of flood risk (2015).

Given the high-level of perceived exposure to flood risk (and actual losses from flooding), significant policy attention has been allocated in recent years to identifying effective means to manage the financial impacts of flooding. As in the case of other natural hazards, governments have a number of tools for managing the financial impacts of flood risk, ranging from investments in risk prevention and public awareness, to the use of risk transfer tools to protect against significant post-disaster costs. A key challenge for governments is determining the most effective and efficient use of public resources for managing disaster risks – in an environment of significant uncertainty that complicates considerably the assessment of flood risk, along with the multiple decisions on prevention, risk reduction, and financial protection that rely on that assessment.

Insurance and other risk financing and transfer tools can make a critical contribution to the financial management of flood risk by spreading disaster risks across domestic and international (re)insurance and capital markets and reducing the share of losses absorbed by households, businesses and governments. However, there are particular challenges to the insurability of flood risk which impedes the availability of affordable private insurance coverage for this peril in many countries, evident in low levels of penetration as well as significant variation in penetration levels across countries. With the exception of flash floods (which can occur anywhere), flood risk is concentrated in well-known locations along rivers and coasts which often face a level of flood frequency (return period) that makes it challenging to charge actuarially-sound insurance premiums that are within the capacity of households to pay. This leads to significant underinsurance of flood risk and leaves governments with difficult decisions on how to best protect vulnerable populations without exacerbating moral hazard or reducing households' incentive to reduce their risk. These challenges have led to a number of reviews and examinations devoted to identifying the appropriate role of government(s) in supporting financial protection against flood risk, and where government intervention is necessary, the most efficient and effective approach.[1]

This report supports governments and policymakers in their efforts to improve the financial management of flood risk and build financial resilience against this risk. It builds on the OECD's analysis and guidance on the development of disaster risk financing strategies, including the *Disaster Risk Assessment and Risk Financing: A G20/OECD Methodological Framework* and the draft Recommendation of the OECD Council on Disaster Risk Financing Strategies which will replace the Recommendation of the OECD Council on Good Practices for Mitigating and Financing Catastrophic Risks (2010).[2] This guidance highlights the critical role of Finance Ministers/Finance Ministries in understanding their country's financial vulnerabilities to disaster risks, based on a comprehensive assessment of financial exposure to disaster risks relative to the capacity to absorb those risks across all segments of society (households, business, local and regional governments, financial sector, etc.).

Chapter 2 provides an overview of the evolving nature of flood risk, including trends in economic impacts over time and the potential implications of climate change. Chapter 3 provides an overview of the challenges in providing financial protection for flood risk across countries, including the extent of underinsurance of flood risk as well as the main challenges to the insurability of flood risk. Chapter 4 outlines possible measures for improving the insurability of flood risk, such as investments in risk reduction and measures to address limited demand for flood insurance. Chapter 5 considers issues related to managing the fiscal costs of floods, including the costs that governments face as a result of flood events and means to minimise and/or transfer those costs. Chapter 6 concludes with some recommendations for designing effective strategies for the financial management of flood risk.

The report is partly based on responses to an OECD survey questionnaire received from 27 countries from across the world: Australia, Austria, Canada, Chile, Costa Rica, Czech Republic, Estonia, France, Hungary, Iceland, Ireland, Israel, Japan, Latvia, Mexico, Myanmar, New Zealand, Peru, Philippines, Poland, Portugal, Russia, Spain,

 

Switzerland, Turkey, United States and Viet Nam. These countries exhibit varying levels of resilience against flood risk and significant diversity in terms of the resources and capacity available to invest in building resilience. Importantly, they also differ substantially in terms of the level of insurance market development which has important implications for the relevance of some of the findings of this report. That said, efforts have been made throughout the report, and particularly in the sections that summarise key findings, to identify where differing country circumstances are important considerations.

## Notes

1. For example, analysis and consultations leading to the establishment of Flood Re in the United Kingdom and the recent announcement of a National Flood Resilience Review, ongoing reviews of the availability of flood insurance in Canada and the Netherlands, various recent legislative actions related to the US National Flood Insurance Program and the findings of the Australian Productivity Commission's review of natural disaster funding arrangements (which was driven by increased losses from numerous disaster events, including floods).

2. As the result of a review of the Recommendation of the OECD Council on Good Practices for Mitigating and Financing Catastrophic Risks (2010), the OECD is developing a Recommendation on Disaster Risk Financing Strategies to replace the original. The draft text for the new Recommendation was made available for public comment until 15 April 2016 (see: *www.oecd.org/pensions/public-consultation-drf.htm*). At the time of writing, a draft Recommendation is being prepared for adoption by OECD Council.

## References


Climate Central (2014), "New Analysis Shows Global Exposure to Sea Level Rise", Climate Central, Princeton, N.J., *www.climatecentral.org/news/new-analysis-global-exposure-to-sea-level-rise-flooding-18066*, accessed 15 October 2015.

Collins, E. and L. Simpson (2007), "The impact of climate change on insuring flood risks", Presented to the Institute of Actuaries of Australia Biennial Convention, 23-26 September, Christchurch, New Zealand, *http://actuaries.asn.au/Library/1.d_Conv07_Paper_Collins%20Simpson_The%20impact%20of%20climate%20change%20on%20insuring%20flood%20risk.pdf.*

Green, C. and E. Penning-Roswell (2004), "Flood Insurance and Government: "Parasitic" and "Symbiotic" Relations", *The Geneva Papers on Risk and Insurance*, Vol. 29 (3), The Geneva Association (The International Association for the Study of Insurance Economics).

Insurance Bureau of Canada (2015), *The financial management of flood risk - an international review: lessons learned from flood management programs in G8 countries*, Insurance Bureau of Canada, Toronto (June).

IPCC (2012), *Managing the Risks of Extreme Events and Disasters to Advance Climate Change Adaptation: A Special Report of Working Groups I and II of the Intergovernmental Panel on Climate Change*, Cambridge University Press, Cambridge, UK, and New York, *http://www.ipcc.ch/report/srex/.*

Jha, Abhas K. and Z. Stanton-Geddes (eds.) (2013), *Strong, Safe, and Resilient: A Strategic Policy Guide for Disaster Risk Management in East Asia and the Pacific*, World Bank, Washington, DC.

Jones-Bos, R. (2011), "As the Mississippi floods, follow the Dutch Model", *The Washington Post*, 26 May, *www.washingtonpost.com/opinions/as-the-mississippi-floods-follow-the-dutch-model/2011/05/23/AGP9kICH_story.html.*

Keating, A. et al. (2014), *Operationalizing Resilience against Natural Disaster Risk: Opportunities, Barriers, and a Way Forward*, Zurich Flood Resilience Alliance.

Kousky, C. and E. Michel-Kerjan (2015), "Examining flood insurance claims in the United States: six key findings", *The Journal of Risk and Insurance (forthcoming).*

Kron, W. (2015), "Global Aspects of Flood Risk Management", *Journal of Hydraulic Engineering*, Vol. 1, pp. 35-46.

Michel-Kerjan, E. and H. Kunreuther (2011), "Disaster Management: Redesigning Flood Insurance", *Science*, Vol. 333, pp. 408-409.

OECD (2012), *Disaster Risk Assessment and Risk Financing: A G20/OECD Methodological Framework*, *www.oecd.org/daf/fin/insurance/g20oecdframeworkfordisasterriskmanagement.htm.*

OECD (2010), Recommendation of the OECD Council on Good Practices for Mitigating and Financing Catastrophic Risks, *www.oecd.org/pensions/insurance/47170156.pdf.*

Productivity Commission (2014), *Natural Disaster Funding Arrangements, Inquiry Report no. 74*, Canberra.

Ramsbottom, D., P. Sayers and M. Panzeri (2012), "Climate Change Risk Assessment for the Floods and Coastal Erosion Sector", report for the UK 2012 Climate Change Risk Assessment, *http://randd.defra.gov.uk/Document.aspx?Document=CCRAfortheFloodsandCoastalErosionSector.pdf.*

Risk Management Solutions and Lloyd's (2008), *Coastal Communities and Climate Change: Maintaining Future Insurability*, Lloyd's, London.

Sadoff, C.W. et al. (2015), *Securing Water, Sustaining Growth: Report of the GWP/OECD Task Force on Water Security and Sustainable Growth*, University of Oxford.

Sato, T. (2006), "Fundamental Characteristics of Flood Risk in Japan's Urban Areas", in *A better integrated management of disaster risks: Toward resilient society to emerging disaster risks in mega-cities*, TERRAPUB and NIED, Tokyo.

Swiss Re (2012), *Flood – an underestimated risk: Inspect, Inform, Insure*, Swiss Reinsurance Company Ltd., Zurich.

Swiss Re (1998), *Floods – an insurable risks*, Swiss Reinsurance Company Ltd., Zurich.

UNISDR (2013), *2013 floods a "turning point"*, UNISDR, Geneva, 25 June, *www.unisdr.org/archive/33693*, accessed 6 January 2016.

*Chapter 2*

# Flood risk in a changing climate

*This chapter of the report provides an overview of trends in the occurrence of floods, including the potential for climate change to impact the frequency and intensity of floods. This is followed by an examination of trends in the economic impact of floods, including direct and indirect losses, disruptions to economic activity as well as the impact of insurance in reducing the economic impacts of natural disasters such as floods.*

Most floods, defined as "the overflowing of the normal confines of a stream or other body of water or the accumulation of water over areas that are not normally submerged" (IPCC, 2012), can be classified into the following categories:

- Flash flood: Heavy or excessive rainfall in a short period of time that, due to the inability for the ground to absorb a high proportion of the water, produces runoff. A flash flood can occur anywhere (usually in conjunction with a thunderstorm or tropical cyclone) and is the most frequent type of flood.
- Riverine flood: Flooding that results from the overflow of water from a stream or river channel onto normally dry land in the floodplain adjacent to the channel. Riverine flooding may occur seasonally as a result of rainy seasons and/or the melting of snow or could occur as a result of abnormally high levels of precipitation that saturates the soil and leads to an increased proportion of rainfall flowing into water courses.
- Coastal flood/storm surge: An abnormal rise in sea level generated by a tropical cyclone or other intense storm that thrusts sea water onto the coast and/or creates large waves as a result of strong winds.
- Ice jam flood: Where flood waves are created by the break-up of ice that had been obstructing the flow of water. Ice jams tend to develop near river bends and obstructions (e.g., bridges).
- Groundwater flood: In locations where the groundwater level is relatively close to the surface, an abnormal increase in rainfall levels can raise the water table leading to damage by water seepage into basements and the destabilisation of building foundations.
- Dam burst: A failure of a dam, resulting from high levels of precipitation, landslides or engineering defects, could lead to significant downstream flooding.
- Debris flows: Where water transports large amounts of solid matter (soil, sand, gravel, rock and/or other debris), including mudflows (i.e. debris flows consisting of small particles). Debris flows can be a combination of landslide and flood and may occur where heavy rain saturates loose soil on a slope. A special form of mudflow is a lahar where rain washes off volcanic ash.

Flooding may also be caused by a tsunami following the displacement of water by an earthquake, volcanic eruption or landslide. However, from the perspective of insurance coverage, damage resulting from a tsunami is usually treated as part of the consequence of the initial event (earthquake, volcanic eruption) and insured accordingly.

Due to their nature, different types of flood pose different types of dangers. Flooding from the sea and large rivers (including as a result of ice jam or dam burst) is generally less frequent although can impact large areas and cause extensive damage. In the case of riverine floods, flood waters generally remain for longer periods of time leading to greater disruption. Sea surge tends to create significant loss potential due to the high-velocity of water inundation and the damaging impacts of salt water. For example, in the United States, individual claims submitted to the National Flood Insurance Program (NFIP) due to storm surge damage have been found to be 8.0% to 20.5% higher than claims from other types of flooding (Kousky and Michel-Kerjan, 2015). Flash floods can cause significant damage due to the more limited advance warning of their occurrence (and therefore more limited time to put in place protective measures) (Kron, 2015).

 

The location where floods occur will also have significant implications for the amount of people affected and the level of damage incurred. Most obviously, the inundation of highly-populated areas will increase the likelihood of large damages. The overall level of damage will also vary with the relative value of assets (floods in developed countries with higher-value assets will tend to lead to larger overall losses than floods in lower income countries). In urban areas, flood water can become polluted with sewage leading to additional health risks and potentially higher clean-up costs (Ramsbottom, Sayers and Panzeri, 2012). Also, variations in altitude within the inundated area can have substantial implications. Relatively flat areas may face longer inundation periods while hilly areas could face higher-velocity water flows with greater potential for damage (Kron, 2015).

## 2.1 Trends in the occurrence and impact of flood events

There has been a substantial increase in the number of flood disasters recorded in recent decades (partly driven by improvements in reporting and data capture). According to data collected by the Centre for Research on the Epidemiology of Disasters (CRED) and included in the EM-DAT database on historical disaster events, close to 62% of the 4 250 flood disasters that have been reported since 1970 have occurred since the year 2000 (see Figure 2.1).[1] The average annual number of reported flood events has increased from under 30 between 1971-80 to approximately 50 between 1981-1990 to over 140 between 2011 and 2015.

Figure 2.1. **Number of flood events by type: 1971-2015 (total number of events during each 5-year period)**

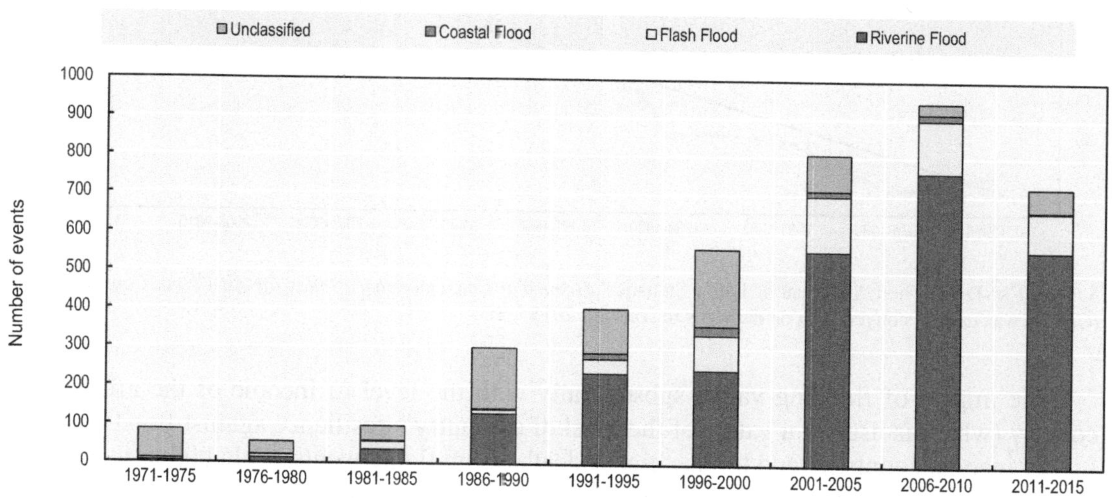


*Source:* EM-DAT.

Since 2000, floods classified as riverine floods have accounted for approximately 73% of all floods, flash floods for close to 16% and coastal floods for just under 2% (the remaining 9% were not classified). However, many flooding events are difficult to classify as an event may involve more than one type of flooding (e.g. a tropical cyclone may cause coastal flooding as a result of wind-driven sea surge, flash flooding due to heavy precipitation accompanying the cyclone and potentially riverine flooding as the accumulated water enters the river system). In addition, flooding resulting from a tropical cyclone will often be classified as part of the meteorological event rather as a separate hydrological event and therefore not recorded as a flood in the EM-DAT data and other data sets.

Annual average damages from floods reported in the EM-DAT database have increased over time, from less than USD 4 billion per year between 1971-1980 (in constant 2015 dollars) to over USD 40 billion per year between 2011 and 2015 (see Figure 2.2). This is consistent with the finding from Kundzewicz et al. (2014) that fluvial flood losses at the global level have increased from approximately USD 7 billion per year during the 1980s to USD 24 billion per year during 2001-2011 (in constant dollars). As noted above, these figures do not generally include flood damages resulting from tropical cyclones which have also increased significantly (from less than USD 6 billion in recorded damages annually between 1971-1980 to over USD 45 billion between 2001-2010 and just under USD 28 billion between 2011 and 2015, in constant 2015 USD and including both damages from wind and flood).

Figure 2.2. **Annual average damage from flood events: 1971-2015 (average annual damage during each 5-year period)**

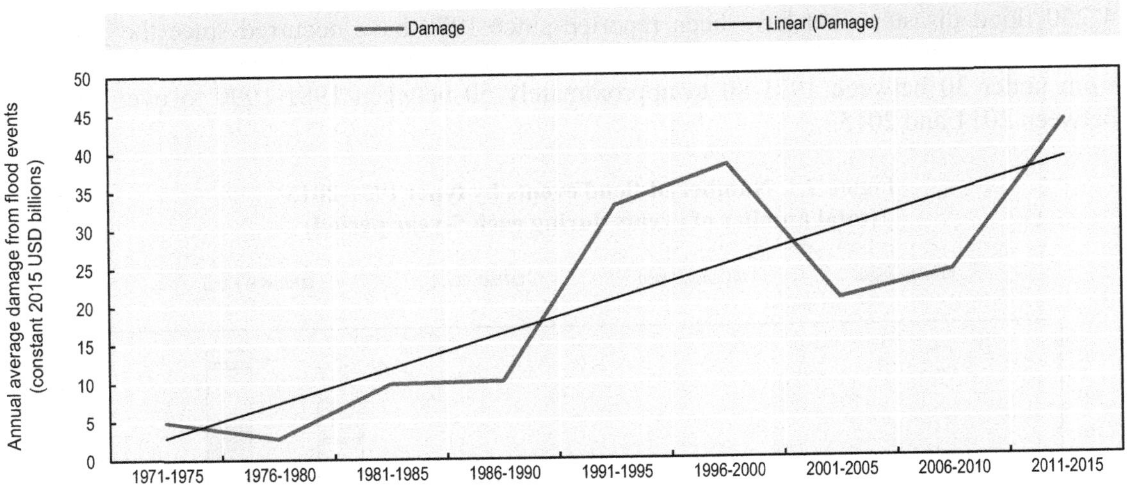


*Source*: EM-DAT. The US Bureau of Labor Statistics' *Historical Consumer Price Index for All Urban Consumers (CPI-U)* was used to convert data on damages to constant 2015 USD.

The impact of flooding varies substantially with the level of income of the affected country (which is usually a gauge of the level of a country's resilience against flood risk). Lower income countries tend to face higher deaths from flood events while higher income countries face higher levels of damage. While 49% of flood events recorded in EM-DAT between 1971 and 2015 have occurred in countries considered low income or lower middle income, more than 60% of all deaths have occurred in those countries. High income and upper middle income countries accounted for just under 80% of all reported damages from flood events (see Figure 2.3).

Figure 2.3. **Flood events, deaths, affected people and damage by income classification**

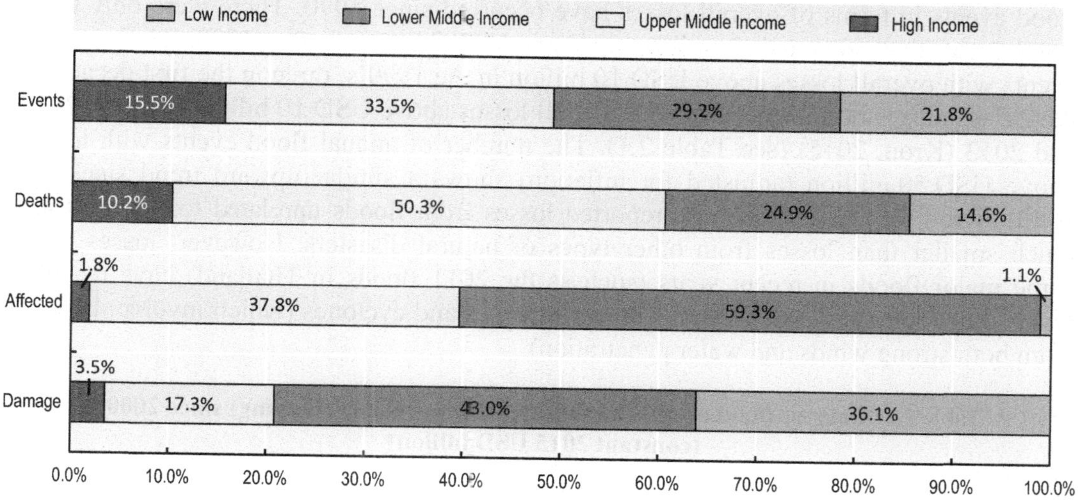


*Source:* EM-DAT. The categorisation of countries by income level was undertaken based on the World Bank FY2016 country and lending groups.

Despite the significant increase in the number of flood disasters, the average annual number of deaths has not increased significantly in recent decades. On average, close to 5 250 deaths were reported annually as a result of flood disasters between 2011 and 2015, relative to an average of 5 800 between 1971 and 1980 (despite the significant increase in reported events). However, the average annual number of people affected per event, and particularly the average damage per event, generally remained at the same level (with significant year-to-year volatility) over the past four decades (i.e. the number of people affected and the level of damage has increased with the number of recorded events - see Figure 2.4). This suggests more significant achievements in terms of protecting people's lives from floods (likely as a result of improved emergency preparedness and response, and in particular early warning capacity) than protecting settlements and property. As suggested by Figure 2.3 above, the average number of deaths and affected people per event is significantly higher in low and middle income countries while damage per event is significantly higher in high income countries.

Figure 2.4. **Average deaths, affected and damage per flood event: 1971-2015**

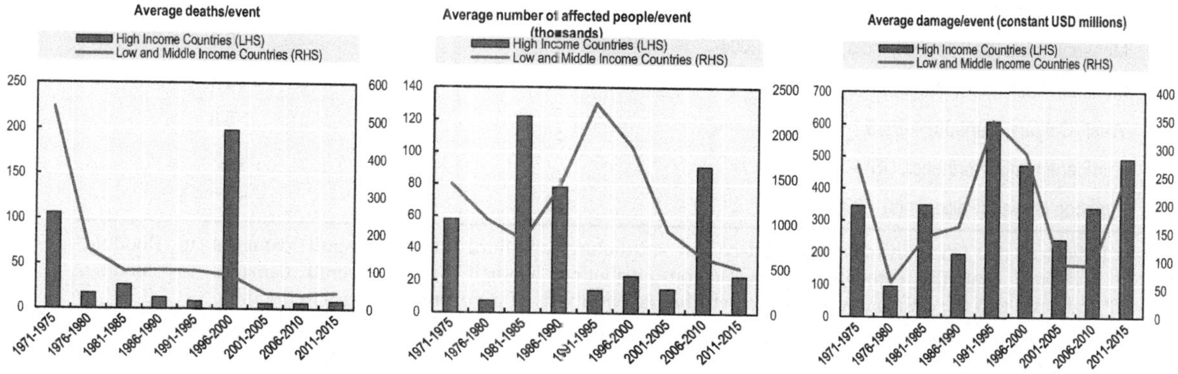


*Source*: EM-DAT. The categorisation of countries by income level was undertaken based on the World Bank FY2016 country and lending groups. The US Bureau of Labor Statistics' *Historical Consumer Price Index for All Urban Consumers (CPI-U)* was used to convert data on damages to constant 2015 USD.

Consistent with the increasing trend in damage from flood events, many of the largest flood events in terms of overall losses have occurred since 1990. There were only three flood events that generated overall losses above USD 2 billion before 1990, relative to six events with overall losses above USD 10 billion in the 1990's, three in the first decade of the 21st century, and three events with overall losses above USD 10 billion between 2010 and 2013 (Kron, 2015) (see Table 2.1). The number of annual flood events with losses above USD 50 million (adjusted for inflation) shows a similar upward trend since the 1980s (Kron, 2015). Historically, reported losses from floods unrelated to cyclones are much smaller than losses from other types of natural disasters. However, losses from some major floods in recent years (such as the 2011 floods in Thailand) have reached levels more commonly associated with earthquakes and cyclones (which involve damage from both strong winds and water penetration).

Table 2.1. **Largest flood events (including cyclone-related flooding) since 2000 (constant 2015 USD billion)**

| Event | Estimated overall losses due to flood |
|---|---|
| Hurricane Katrina (US Gulf Coast) – 2005 | 100.7*** |
| Hurricane Sandy (US Northeast) – 2012 | 47.5*** |
| Chao Phraya (Thailand ) – 2011 | 45.3 |
| Elbe/Danube (Central and Southern Europe) – 2002 | 21.7 |
| Hurricane Ike (Caribbean, US) – 2008 | 14.3* |
| Elbe/Danube (Central Europe) – 2013 | 12.8 |
| Southern Alps (Italy and Switzerland) – 2000 | 11.7 |
| Midwest/Missouri (US) – 2008 | 11.0 |
| Indus (Pakistan) – 2010 | 10.3 |
| Centre, South, East, Northwest (China) – 2003 | 10.2 |
| Hurricane Ivan (Caribbean, US) – 2004 | 10.0* |
| Southwest, Centre, Northwest (China) – 2004 | 9.8 |
| East, Southeast, South (China) – 2010 | 8.7 |
| Hurricane Wilma (Caribbean, Mexico, US) – 2005 | 8.5* |
| East, Northeast, Southeast (China) – 2012 | 8.3 |
| Tropical Storm Allison (Houston, US) – 2001 | 8.0 |
| South, Southwest, East, Centre (China) – 2007 | 7.8 |
| Monsoon rains (Bangladesh, India, Nepal) – 2004 | 6.3 |
| Monsoon flash flood (Mumbai, India) – 2005 | 6.1 |
| West (Calgary, Canada) – 2013 | 5.8 |
| Hurricane Irene (Northeast, US) – 2011 | 5.3** |
| Typhoon Haiyan (Philippines) – 2013 | 5.1** |

* Indicates that the estimate was based on attributing one-third of the overall damages to flooding.
** Indicates that the estimate was based on attributing one-half of the overall damages to flooding.
*** Indicates that the estimate was based on attributing two-thirds of the overall damages to flooding.

*Source*: The list of events, including estimates of overall losses at original value and the share of overall damage due to flooding, are taken from Kron (2015) using data from Munich Re's NATCATSERVICE. The US Bureau of Labor Statistics' *Historical Consumer Price Index for All Urban Consumers (CPI-U)* was used to convert data on damages to constant 2015 USD.

## 2.2 The economic impact of floods

Direct losses from floods are significant for many countries. According to the EM-DAT data, since 1990, 36 countries have faced at least one year of damages to property, crops and livestock of USD 1 billion (in constant 2015 USD) or more from floods while 15 countries experienced at least one year of flood damages exceeding USD 5 billion (including Australia, Bangladesh, Canada, China, Germany, India, Iran, Italy, Japan, North Korea, Pakistan, Poland, Thailand, United Kingdom, and the United States). In some countries, annual average recorded damages have accounted for a material share of GDP (see Figure 2.5). A 100-year event in Central Europe could cause property damages of 2.5% of GDP in the Czech Republic; 3.2% in Poland; 4.6% in Hungary; and 8.5% in the Slovak Republic (Pollner, 2012). In some countries, including the United States, China and India, expected annual damages (EAD) of more than USD 10 billion have been estimated by some analyses (Sadoff et al., 2015).

Figure 2.5. **Annual average damage from flood events as a share of GDP**

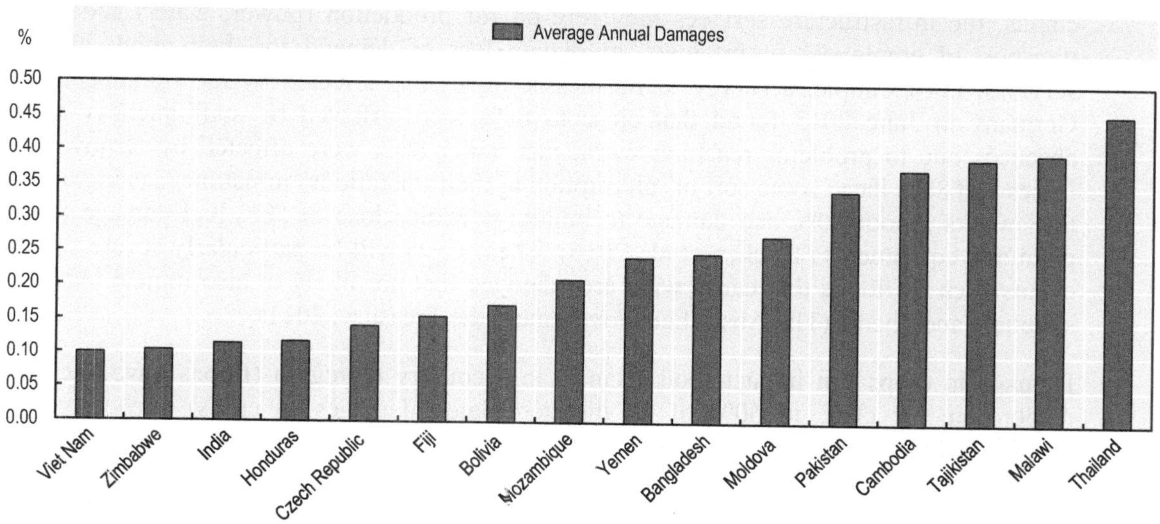


*Notes:* Annual average damage was calculated based on damage reported between 1971 and 2015 and converted to constant 2015 USD based on the US Bureau of Labor Statistics' *Historical Consumer Price Index for All Urban Consumers (CPI-U)*. GDP figures are from the World Bank for the year 2014 (most recent year available) at current USD (*http://data.worldbank.org/indicator/NY.GDP.MKTP.CD*).

*Source*: EM-DAT.

A key driver of the increasing occurrence and losses from floods and cyclones is the accumulation of assets in flood-prone areas (a flood or cyclone would only be reported as a disaster if it affects a populated area). For example, in the United States, the population in counties along the Atlantic, Pacific, and Gulf Coasts has increased from approximately 47 million in 1960 to 87 million people in 2008 and the number of housing units along the coast has more than doubled from 16.1 million in 1960 to 36.3 million in 2008 (Wilson and Fischetti, 2010).

Based on the accumulation of assets alone (i.e. not taking into account changes in the nature of flood occurrence), exposure to flood risk is expected to increase significantly. While global projections are challenging because of differences in data availability and quality, a number of studies have attempted to project the level of exposure in future years based on asset accumulation:

- Jongman, Ward and Aerts (2012) estimate that the value of global assets exposed to 1-in-100 year river flooding will increase by 200-250% between 2010 and 2050 (depending on the calculation method used) while the value of assets exposed to 1-in-100 year coastal flooding will increase by 182-200%.[2]

- According to estimates by Güneralp et al. (2015) based on urban land-use projections, 13% of urban land will be located in "low-elevation coastal zones"[3] vulnerable to flooding and 40% of urban land will be located in high-frequency flood zones by 2030 (from 11% and 30%, respectively in 2000), with developing countries accounting for an increasing share of that exposure as a result of more rapid urbanisation.

Large flood events can also have significant financial and economic implications for government, business and households through indirect consequences such as business interruption,[4] loss of employment and output, and decreased tax revenues (as well as significant social and environmental consequences):

- Indirect impacts on businesses may occur as a result of disruptions to their supply chains, the infrastructure services they rely on for production (power, water) and/or transport of employees or products, or due to loss of demand for their goods and services. For example, a survey of businesses in regions affected by the flooding in Germany in June 2013 found that close to 60% were affected by staff lateness or absences due to problems reaching work, just under 80% were affected by turnover losses and 88% faced some sort of interruption to their operations, sometimes lasting up to 8 weeks (about 80% had damage to buildings while close to 50% had damage to plant and equipment) (Thieken et al., 2016). These costs will be particularly severe for events such as floods that cause wide-area damage that are likely to also impact local suppliers and clients (The Australian Government the Treasury, 2011).

- Damage to crops can impact food security in a country or region (floods have been responsible for close to 60% of all disaster-related damages to crops). Repetitive flooding could exacerbate food crises over several years and even have an impact on international food markets if a major producer country is affected (FAO, 2012).

- Tourism revenues may also be significantly impacted if the flood event changes perceptions about the attractiveness or safety of a disaster-affected country/city. For example, the 2011 floods in Thailand led to estimated losses in tourism revenue of USD 3 billion (World Bank and Thai Ministry of Finance, 2012).

At the macro-level, government finances could be impacted by the loss of corporate tax revenue due to business interruption and personal income taxes due to lost wages. Governments, both local and national, are also likely to face significant costs related to recovery and reconstruction (often included as an indirect economic cost of disasters). For example, tax revenues in New York City were estimated to have declined by USD 160 million after Hurricane Sandy due to lost business revenue and wages (New York City Recovery, 2015). Governments may also face an adverse impact on balance of payments if exports or capital flows are significantly affected by the disaster event.

While indirect costs are difficult to calculate, a number of pre- and post-event studies have attempted to estimate these costs. For example, an OECD (2014) study assessing the impacts of a major flood event in the Paris region estimated the magnitude of a number of indirect costs as a result of disruptions to critical infrastructure (Box 2.1).

### Box 2.1. **Indirect economic impacts: Seine river flood in Île-de-France**

In 2014, the OECD undertook an assessment of the potential economic impacts of a major flood of the Seine river in the greater Paris region. The assessment considered three flood scenarios based on historical occurrences and used business surveys and economic modelling to estimate the potential direct, indirect and overall macro-economic impacts. Based on input from the operators of major power, transport and water utilities, the study estimated the scope of disruption to critical infrastructure services:

- Based on the location of sub-stations, in an extreme scenario (1910 flood levels), approximately 1.5 million households and business customers could face power supply disruptions over an area 50% larger than the area affected directly by floods.
- More than half of the 250 km metro line would be closed, leaving only one of the 14 Paris metro lines operational. A number of suburban public transport lines would also be disrupted along with 3 major train stations in Paris. In addition, the road network would be significantly disrupted, including five motorways, several major highways and all bridge crossing across the Seine River (which winds through Paris and its suburban region).
- The drinking water supply could be disrupted in the outskirts of Paris, with more than 5 million customers potentially facing extended water supply disruptions and 1.3 million customers facing deterioration in water quality.

The impact of these disruptions on businesses' operating losses (particularly as a result of power and transport disruptions) were estimated at EUR 19 billion, or almost 65% of the direct losses of EUR 29.4 billion estimated for the most extreme scenario.

*Source*: OECD (2014)

The globalisation of supply chains means such disruptions can also have regional or even global impacts. For example, the flooding of several industrial parks in Thailand in 2011 had global/regional impacts in many sectors (including automotive and electronics) as global companies such as Toyota, Honda, Nissan, Ford, Apple, Sony, Canon, and Toshiba faced disruptions to production or facility closures as a result of their linkages to sites located in the flood zone (Chachavalpongpun, 2011). Global industrial production declined by 2.5% as a result of the floods (Schanz and Wang, 2015). The 2015 flooding in Chennai, a major automotive production centre in India, affected an estimated 10-15% of India's automotive production as a result of plant shutdowns and supply chain impacts (Thakkar, 2015).

These direct and indirect losses can have a significant impact on the broader economy. Von Peter, von Dahlen and Saxena (2012), using data from Munich Re's NatCatSERVICE for 2 476 major natural catastrophes in 203 countries between 1960-2011, found that the average natural disaster (of all types, including floods) leads to a fall in growth of 1.0% of GDP upon impact and a cumulative loss to GDP of 2.6%. These impacts are particularly severe for developing countries, and the poorest households within those countries, due to their more limited capacity to manage disaster risks. The Lloyd's City Risk Index: 2015-2025 (2015) also provides an estimate of the economic output at risk from various perils in 301 major cities, including floods. According to Lloyd's, USD 432 billion of economic output is at risk[5] from coastal and riverine flooding in the 301 cities analysed, including more than USD 10 billion in each of the seven most exposed cities (Tokyo, Osaka, Los Angeles, New York, Sao Paulo, Delhi and Chinese Taipei).

## 2.3 Potential impact of climate change on the intensity and frequency of flood events

While the evidence is far from conclusive, climate change is expected to impact the nature of flood risk going forward as a result of changes to: i) the frequency of heavy precipitation events; ii) the range and intensity of cyclones; and iii) the rise in sea-levels. Specifically, in their special report on *Managing the risks of extreme events and disasters to advance climate change adaptation* (2012),[6] the Inter-Governmental Panel on Climate Change (IPCC) found evidence of a number of likely impacts of climate change on the nature of flood events (although subject to significant regional variation and various levels of uncertainty), including:

- the frequency of heavy precipitation or the proportion of total rainfall from heavy rainfalls will likely increase over many areas of the globe as higher air temperatures allow the atmosphere to retain more water;
- the average tropical cyclone maximum wind speed will likely increase in some ocean basins;
- there may be a projected poleward shift of extratropical storm tracks;
- the mean sea level rise will very likely contribute to upward trends in extreme coastal high water levels; and
- changes in heat waves, glacial retreat, and/or permafrost degradation will affect high mountain phenomena such as slope instabilities, movements of mass, and glacial lake outburst floods.

Among these potential impacts, the IPCC places more confidence in the predicted impact on rainfall intensities than other natural disasters (IPCC, 2012). An increase in the occurrence of heavy precipitation events could increase the frequency of flash floods, riverine floods and groundwater floods. Increasing levels of urbanisation, which will generally reduce the water absorption capacity of land by converting natural terrain to urban use, will likely exacerbate these climate change impacts and increase the level of resulting damage (Wilby and Keenan, 2012).

While there is significant uncertainty in assessing the potential implications of climate change on flood risk, a number of studies have analysed this issue. For example, an assessment of likely changes in river flooding return periods (Arnell et al., 2014) found that the frequency of river flooding will likely double (or more) by 2050 (relative to the period 1961-1990) in Central and Eastern Europe, Central America, Brazil and some parts of Western and Central Africa – while decreases in frequency can be expected in some parts of Asia. Based on the projections of numerous climate models, the return periods for what are currently considered 1-in-100 year floods is expected to decline (i.e. occur more frequently) by 2100 in 22 of the 29 major river basins examined, and decline significantly in a number of basins including the Lena (Northeast Eurasia), Congo (Central Africa), Nile (East Africa), Ganges (South Asia), Mekong (Southeast Asia), and Murray (Australia) (Hirabayashi et al., 2013). Close to 90% of the respondents to an OECD questionnaire on the financial management of flood risk indicated that climate change has increased the level of flood risk in their country, with almost 50% indicating that that impact has already been significant.

A number of studies have used climate change scenarios to model the possible impacts of climate change on flood losses for different regions of the world (for various time periods), using a range of approaches to estimating the damage that could occur as a result of predicted changes in weather patterns. Box 2.2 provides an overview of selected studies that have been undertaken for inland flooding and the range of estimates derived.

Sea-level rise, which is also predicted with a higher-level of confidence, increases the risk of coastal flooding. A study of communities along the US Eastern and Gulf coasts found that, even under a mid-range scenario for future sea level rise, two-thirds of communities could face an increase in the frequency of high-tide flooding (i.e. tidal flooding under normal (non-storm) conditions) of 300% from current levels, with a number of communities facing regular extensive flooding from high-tides alone (Spanger-Siegfried, Fitzpatrick and Dahl, 2014). A study analysing 55 tidal gauges across the United States estimated that the occurrence of what are currently considered 1-in-100 year high water levels could increase to 1-in-10 years in many communities by 2050, and events that are currently considered 1-in-10 year events could occur annually (Tebaldi, Strauss and Zervas, 2012). Depending on the level of sea-level rise, coastal flooding in some vulnerable small island states could become pervasive. The occurrence of coastal flooding could also increase as a result of the higher levels of precipitation that would accompany more intense cyclones (increases of approximately 20% in the precipitation rate within 100 km of the storm centre are generally predicted in higher resolution modelling studies (Knutson et al., 2010)).

### Box 2.2. **The potential impact of climate change on losses from inland flooding**

Climate change-related increases in the risk of inland flooding may occur as a result of increased precipitation, more frequent heavy precipitation events, or changes in snowmelt patterns. Inland, these changes could lead to an increase in riverine flooding (as a result of the more frequent or intense precipitation or higher levels of snowmelt) and flash flooding (as a result of heavier precipitation, potentially exacerbated by changes in overall levels of precipitation). Estimates of the potential impact of changes to weather patterns on losses have been undertaken for a number of OECD countries and regions, including the United Kingdom (Dailey et al., 2009), France (Moncoulon and Veysseire, 2015), Germany (GDV, 2011 and Te Linde et al., 2011), Spain (Feyen, Barredo and Dankers, 2009), Australia (Schneider et al., 2000), Netherlands (Bouwer, Bubeck and Aerts, 2010 and Hoes, 2007), Canada (Cheng et al., 2012), Norway (Haug et al., 2011) and Europe (Jongman et al., 2014 and Feyen, Barredo and Dankers, 2009).

While these studies use various climate change scenarios, damage calculations, time periods and loss-types (e.g. insured vs. total damage), the results provide a range of estimates of the magnitude of change in losses that could arise as a result of changes in precipitation and snowmelt patterns across a number of countries.[1] By converting these estimates into estimates of annual increases in damage, a comparable range of estimates of climate-change driven impacts can be calculated in order to derive a rough extrapolation of the magnitude of future losses resulting from changes in climate.

Figure 2.6 uses the derived estimate of the annual increase in losses due to climate change to provide a rough extrapolation of global flash flood and riverine flood losses to 2070. The estimates are derived by calculating the level of losses in each year under two scenarios: i) the increase in losses assuming that the annual average increase in losses that occurred between 1990 and 2015[2] is maintained (which would include both the growth in exposed assets as well as any change in the frequency or intensity of flooding); and ii) the additional increase in average annual losses due to changing climate patterns (based on the range of estimates in the studies noted above). As can be seen in the figure, climate change could have a significant impact on overall level of losses, increasing total losses in 2070 by over USD 10 billion relative to estimated damages based on the current trend in average annual increase in losses.

## Box 2.2. The potential impact of climate change on losses from inland flooding (cont.)

### Figure 2.6. Projected annual losses from riverine and flash flood in OECD countries: Impact of climate change

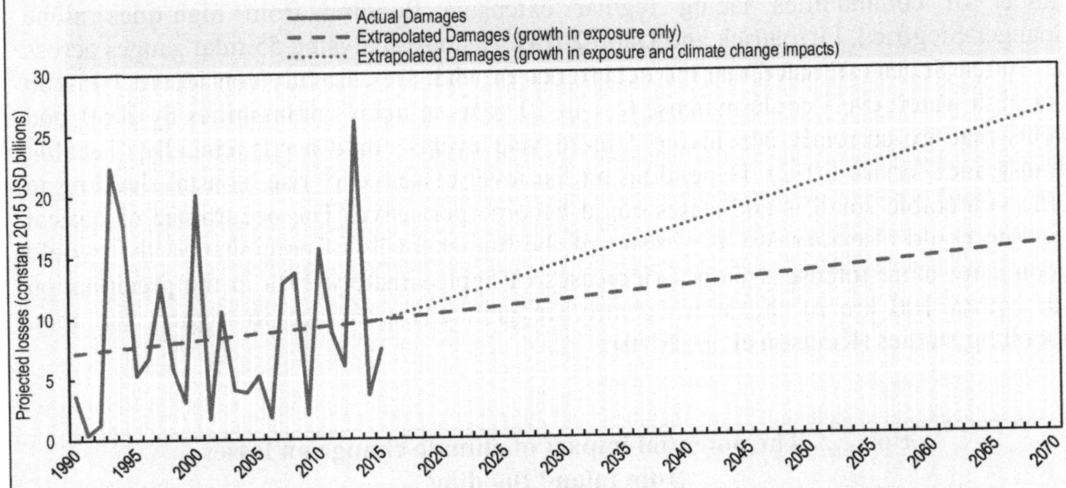


1. While some of the studies' estimates of change include projections of changes in the value of insured assets (or other indicators of socio-economic developments), those that do also provide estimates of the relative share of the change in losses resulting purely from changes in climate.
2. The annual increase in losses from flash floods and riverine floods was calculated based on the increase in losses for both types of flooding in high-income OECD countries as reported in the EM-DAT data since 1990 (approximately 1.5% of 1990 losses or USD 105 million per year in constant 2015 USD). The average annual increase in losses due to climate change impacts was calculated as USD 199 million per year in constant 2015 USD based on the average increase estimated in the studies noted above.
3. The studies used for calculating climate change impacts modelled the following types of flooding: United Kingdom (riverine, coastal and flash flooding in England and Wales and riverine and flash flooding in Scotland and Northern Ireland); France (overland and river runoff flooding in metropolitan France for all river basins); Germany (riverine flooding in 5 major river basins and riverine flooding in the Rhine basin); Spain (riverine flooding in Madrid); Australia (local flooding); Netherlands (riverine flooding and local flooding); Canada (flash flooding in 4 cities in Ontario); Norway (flash flooding in 3 counties); Europe (riverine flooding covering over 1 000 basins and riverine flooding).

*Source*: OECD calculations based on EM-DAT data and sources identified above on the impact of climate change on losses. The US Bureau of Labor Statistics' *Historical Consumer Price Index for All Urban Consumers (CPI-U)* was used to convert data on damages to constant 2015 USD.

An analysis by Standard and Poor's Rating Services (2015) and Swiss Re of the increase in damage (as a share of asset values) from a 1-in-250 year flood event under climate change (relative to no change in the nature of a 1-in-250 year event) found that, on average, climate change would increase damage-to-value ratio by 25% from such an event in the sampled countries by 2050 with some countries facing a potentially significant increase in damage (see Figure 2.7).

## Box 2.2. The potential impact of climate change on losses from inland flooding (cont.)

Figure 2.7. **Damage-to-value ratio from a 1-in-250 year flood: Impact of climate change**

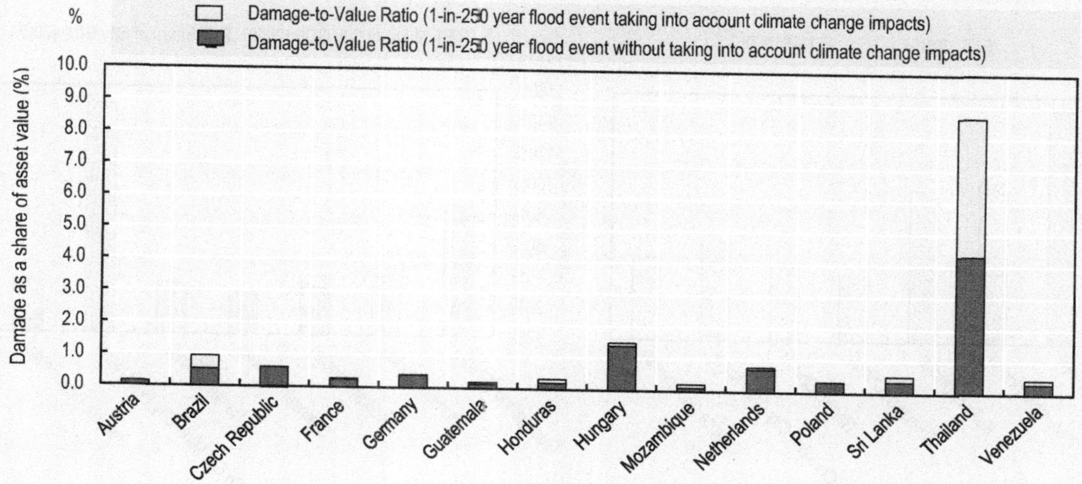


*Source*: Standard and Poor's Ratings Services (2015).

By elevating the base-height of coastal water levels, sea-level rise also increases the possibility of damaging storm surge (RMS and Lloyd's, 2008). When combined with the expectation of higher maximum cyclone wind speeds, the potential for damage is significantly increased. Estimates by RMS (2015) suggest that the probability of events causing at least USD 10 billion in storm surge losses will increase significantly by 2100 as a result of sea-level rise in a number of US coastal cities:

- From 2.22% to 5.26% in Tampa, Florida (1-in-45 years to 1-in-19 years);
- From 1.28% to 5.88% in Miami, Florida (1-in-78 years to 1-in-17 years); and
- From 0.87% to 3.70% in New York, New York (1-in-115 years to 1-in-27 years).

When combined with the expected increase in wind damage, the probability of a USD 10 billion loss event is projected to increase by 1.4 times in Miami and 2.5 times in New York by 2100 (RMS, 2015).

Increases in storm surge losses are also projected for other parts of the world. For example, the IPCC-derived climate change scenario involving a 0.37m sea-level rise in the North Sea could lead to an increase in average annual expected losses from EUR 0.6 billion in 2009 to EUR 2.6 billion in 2100 from storms and sea surges in Northern European countries, with the increase ranging from 100% to 900% (depending on the country) (Swiss Re, 2009). An analysis of the potential impacts of climate change (0.5 m sea-level rise, 10% increase in extreme storm-related water levels and subsidence) as well as asset accumulation in the world's largest port cities projected a more than ten-fold increase in the value of exposed assets, of which approximately 35% of the increase in value was attributable to climate change factors (Hanson et al., 2011).

In Asia, a number of mega-cities are located in coastal areas and are expected to face substantial growth in potential losses as a share of city GDP as a result of population growth and economic development, sea-level rise and subsidence (see Figure 2.8).

Figure 2.8. **One-in-100 Year Flood Exposure in Asian Mega-Cities: 2005 and 2050**

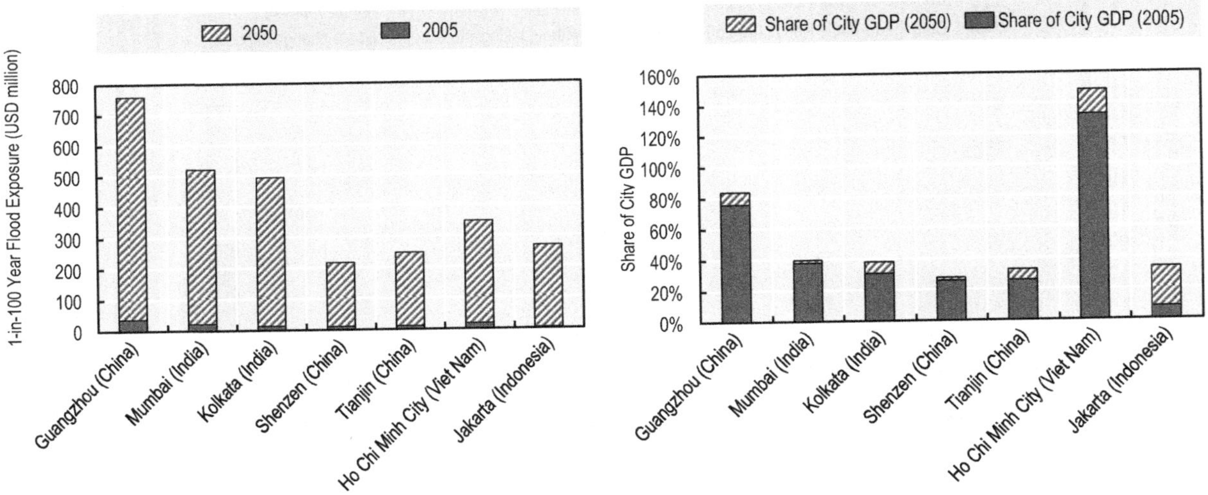


*Source*: OECD calculations based on Schanz and Wang (2015) and Hallegatte et al. (2013).

Some countries have undertaken comprehensive assessments of the potential impact of climate change on all types of flooding. For example, in the United Kingdom, periodic assessments of various sectors, including coastal erosion and flooding, are undertaken to estimate possible changes in the level of risk (see Box 2.3).

## 2.4 The potential role of insurance in reducing economic disruption

The level of insurance penetration has been shown to be negatively correlated with the level of impact of disasters on economic output (i.e. countries with higher levels of insurance penetration face more limited negative impacts on economic output). Using data for high and middle-income countries between 1975 and 2008, Melecky and Raddatz (2011) estimate the impact of geological, climatic, and other types of natural disasters on government expenditures and revenues. They found that countries with lower levels of insurance penetration faced larger declines in economic output and more considerable increases in fiscal deficits in response to disasters than countries with higher levels of insurance penetration. Similarly, von Peter, von Dahlen and Saxena (2012) also provide an estimate for the relative impact of disaster events where losses are insured relative to events with no insurance coverage. They estimate separately the impact of disaster-related losses with and without subsequent insurance payout. They find that insured losses have no statistically significant impact on long-term output (i.e. GDP growth does not diverge significantly from its pre-disaster trend) while uninsured losses come with additional macroeconomic costs, amounting to a cumulative output cost over 10 years of 2.3% or more (see Figure 2.9).

### Box 2.3. **UK climate change risk assessment**

The United Kingdom undertakes climate change risk assessments every five years. As part of the most recent (2012) Climate Change Risk Assessment (CCRA) process, a comprehensive assessment of the potential impacts of climate change on coastal erosion and flooding was undertaken. The assessment was based on UK climate projections ("UKCP09") which predict increases in: i) the rate of sea-level rise, leading to increased coastal flood risk along the coast and in estuaries; and ii) winter precipitation (increase of 12% to 30%) and storm rainfall intensity (doubling of frequency of heavy rainfall events), which would lead to an increase in riverine and flash flooding. The assessment used future climate scenarios and projections of socio-economic changes (i.e. predictions of changes to the level of exposed assets based on population growth and asset accumulation in zones subject to flood risk) to estimate the change in potential exposure to coastal and river flooding (suitable data for analysis of flash flooding was not available, although the assessment suggested that such an analysis would likely show increasing risk).

The assessment provided estimates of the change in the number of people and properties in England and Wales that could be exposed to a high risk of flooding (more than 1-in-75 year return period or annual probability of 1.3% or higher) in the 2050s and 2080s relative to 2009:

- an increase in the number of people at high risk of flooding from 900 000 in 2009 to between 1.7 million and 5 million in the 2080s
- an increase in the number of properties at high risk of flooding from about 560 000 (370 000 residential and 190 000 non-residential) to between 1.0 million and 2.9 million by the 2080s (of which between 700 000 and 2.1 million are residential properties)
- an increase in the Expected Annual Damage (EAD) to properties from GBP 1.2 billion (of which GBP 640 million is the EAD to residential properties) in 2009 to between GBP 2.1 billion and GBP 12 billion in the 2080s (of which GBP 1.2 billion to GBP 6.5 billion is the EAD to residential properties)
- an increase in average annual business interruption costs from GBP 20 million to GBP 60 million by the 2080s
- an increase in average annual insurance payouts from between GBP 200 million and GBP 300 million in 2009 to between GBP 500 million and GBP 1 billion in the 2080s.

The assessment noted that a higher increase in sea level rise ("a plausible low likelihood high impact scenario") could double the number of properties at high risk of coastal flooding. The authors estimated that approximately 60% of the increase (in EAD) to 2080 resulted from the climate change-related increase in flood risk and 40% was due to socio-economic changes.

*Source*: Ramsbottom et al. (2012).

Figure 2.9. **Insurance penetration and the economic impact of disasters**

The role of risk transfer

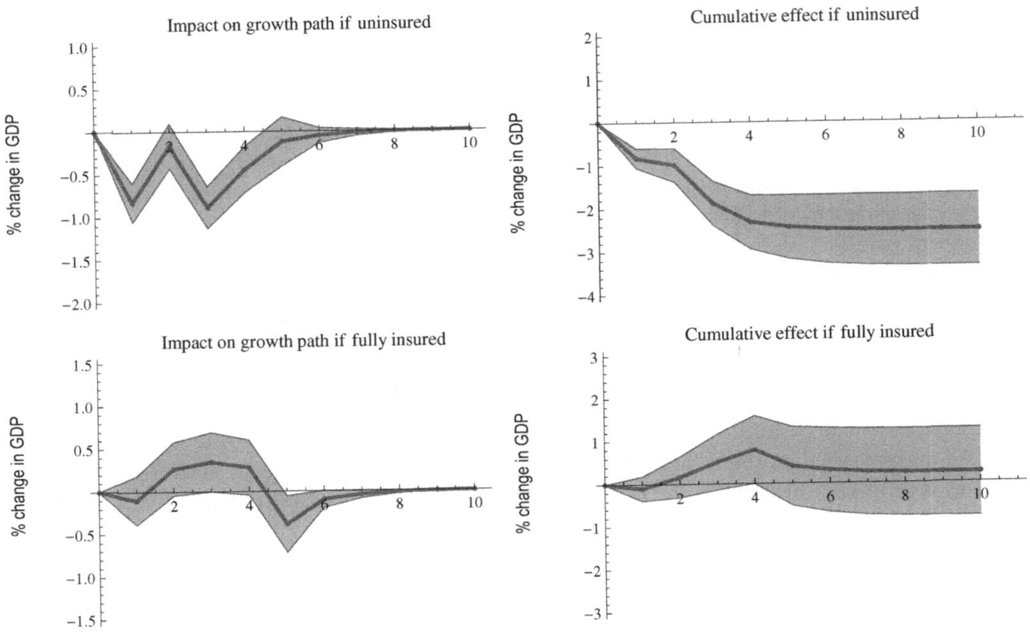


1. The left-side panels show the deviation of real economic growth from its trend due to a typical disaster event for an economy where all losses are uninsured and an economy where all losses are insured. The right-side panels show the cumulative deviation of GDP from trend over 10 years for each type of economy.

*Source:* von Peter, von Dahlen and Saxena (2012).

While the level of insurance penetration is usually higher in high-income countries (with higher levels of overall resilience), there are a number of ways in which insurance might specifically make a positive contribution to reducing economic disruption. Insurance claim payments can provide a timely source of financing for reconstruction (Keating et al., 2014) – a factor that is beginning to be recognised in credit rating agency assessments of sovereign ratings (Standard and Poor's Ratings Services, 2015). A survey of households affected by Hurricane Katrina found that close to 80% of residences that were insured were rebuilt in subsequent years while less than 50% of uninsured properties were rebuilt (Turnham et al., 2011). Insurance payments also tend to be larger and more quickly disbursed than government assistance (Kousky and Shabman, 2015). After flooding in Germany, Austria and Switzerland in 2005, the average times for a claim to be approved by private insurers in Germany and the (public) cantonal monopole insurers in Switzerland were significantly lower than the amount of time taken to approve a claim through the Austrian public compensation fund (21 and 38 days in Switzerland and Germany vs. 53 days in Austria) (Schwarze et al., 2011).

The global nature of international reinsurance markets also means that a portion of the financing of (reinsured) claims payments is likely to be absorbed by international markets and will therefore reduce the burden on national economies. For example, in New Zealand, where earthquake losses are mostly covered by the Earthquake Commission and private insurers (which are, in turn, reinsured in international markets), the economic impact of the 2011 Canterbury earthquake sequence in New Zealand was minimal despite direct losses approaching 20% of GDP (New Zealand Parliamentary Library, 2014).

Another potential contribution could be the reduced cost to taxpayers in countries with high levels of insurance penetration. A Lloyd's (2012) case study of five disasters (US hurricanes in 2005, UK flooding in 2007, Sichuan earthquake in 2008, Great East Japan Earthquake in 2011 and Thailand floods in 2011) found that a larger share of uninsured losses tended to be correlated with a larger overall cost to taxpayers. This is likely because governments faced with significant uninsured private losses after a disaster will generally face political pressure to compensate those affected, leading to negative impacts on public finances (in cases where that compensation was not previously accounted for in public accounts).

Where new taxes have been imposed to fund reconstruction, there may be negative implications on consumption and therefore economic recovery. In addition, in countries where homeowners or businesses maintain low levels of insurance protection against floods (or alternatively, government compensation for flood losses is low), a significant flood event could lead to an increase in defaults on mortgages, other consumer loans and/or commercial loans if debtors are faced with direct or indirect losses that are beyond their capacity to absorb. If such a scenario were to impact credit conditions, it could also be expected to have negative implications for the broader economy.[7]

## Notes

1. The EM-DAT database, maintained by the Centre for Research on the Epidemiology of Disasters, provides data on deaths, affected people, damages and other variables for natural and man-made disasters in all countries since 1900. To be included in the EM-DAT database, a disaster must meet at least one of the following criteria: i) ten or more people reported killed; ii) one hundred or more people reported affected; iii) a declaration of a state of emergency; or iv) a call for international assistance.

2. The authors use a generalised calculation of maximum exposure to damage based on average maximum damage levels per km2 of urban land area (land-use method) and as a function of population (population method). The resulting levels of estimated exposure to damage differ substantially between the two methods although the calculated levels of growth are relatively consistent.

3. For the purposes of the analysis, low-elevation coastal zones (LECZ) are defined as "the contiguous area along the coast that is less than 10 m above sea level" (based on (McGranahan, Balk and Anderson, 2007).

4. Business interruption that occurs as a result of direct damages to structures or equipment involved in production is generally considered part of the direct economic losses of a disaster event.

5. To calculate economic output at risk (or "GDP at risk"), the analysis looks at the potential in lost output over a five-year period after a flood event relative to the baseline (no event) (Cambridge Centre for Risk Studies, 2015).

6. The contribution of Working Group II to the Fifth Assessment Report of the Intergovernmental Panel on Climate Change reconfirmed these general findings and provided further detail on the near-term impacts of climate change (IPCC, 2014).

7. This issue was considered as part of the Natural Disaster Insurance Review in Australia and was the subject of a response by the Australian Bankers Association (ABA). The ABA noted that past disasters had not significantly impacted banks' lending books (negligible losses), that banks are protected because most of the property value was related to the land not the building, that government compensation provides additional protection against defaults, and that banks have a number of other options for protecting against underinsurance of natural disaster risks by homeowners and businesses, including mortgage insurance and higher loan-to-value ratios (some banks did in fact reduce their maximum LTVs in flood prone areas after 2011 floods) (The Australian Government the Treasury, 2011).

## References

Aerts, J., and W. Botzen (2011), "Climate change impacts on pricing long-term flood insurance: A comprehensive study for the Netherlands", *Global Environmental Change*, 21. pp. 1045-1060, *http://doi:10.1016/j.gloenvcha.2011.04.005*.

AIR Worldwide and UK Met Office (2009), "The Financial Risks of Climate Change," *ABI Research Paper No. 19, 2009*, Association of British Insurers, London.

Arnell, N.W. et al. (2014), "The impacts of climate change across the globe: A multisectoral assessment", *Climactic Change*, 11 November.

The Australian Government the Treasury (2011), *Natural Disaster Insurance Review: Inquiry into flood insurance and related matters (September 2011)*, Commonwealth of Australia, *www.ndir.gov.au/content/report/downloads/NDIR_final.pdf*.

Bouwer, L.M., P. Bubeck and J.C.J.H. Aerts (2010), "Changes in future flood risk due to climate and development in a Dutch polder area", *Global Environmental Change*, Vol. 20, pp. 463–471.

Bureau of Labor Statistics (2016), "Table 24: Historical Consumer Price Index for All Urban Consumers (CPI-U)", *CPI Detailed Report (tables 1-29 only) February 2016*, United States Bureau of Labour Statistics, Washington, DC, www.bls.gov/cpi/cpid1602.pdf

Cambridge Centre for Risk Studies (2015), *World Cities Risk 2015-2025: Part II – Methodology Documentation*, Cambridge Centre for Risk Studies, University of Cambridge Judge Business School, Cambridge, United Kingdom.

Chachavalpongpun, P. (2011), "Floods Threaten Global Trade Hub", *Asia Sentinel*, 26 October, *www.asiasentinel.com/econ-business/floods-threaten-global-trade-hub/*, accessed 13 October 2015.

Cheng, C.S. et al. (2012), "Climate change and heavy rainfall-related water damage insurance claims and losses in Ontario, Canada", *Journal of Water Resource and Protection*, 4(2), pp. 49-62.

Dailey, P. et al. (2009), "The Financial Risks of Climate Change: Examining the Financial Implications of Climate Change using Climate Models and Insurance Catastrophe Risk Models", *ABI Research Paper No. 19*, The Association of British Insurers, London.

 

FAO (2012), *The impact of disasters on agriculture and food security*, Food and Agriculture Organisation, Rome.

Feyen, L., J.I. Barredo and R. Dankers (2009), "Implications of global warming and urban land use change on flooding in Europe: Water and urban development paradigms—towards an integration of engineering" in *Design and management approaches*, : J. Feyen, K. Shannon, and M. Neville (eds.), Boca Raton, FL: CRC Press.

GDV (2011), *Auswirkungen des Klimawandels auf die Schadensituation in der deutschen Versicherungswirtschaft. Kurzfassung Hochwasser. Studie im Auftrag des Gesamtverbandes der Deutschen Versicherungswirtschaft e.V.*, German Insurance Association (Gesamtverband der Deutschen Versicherungswirtschaft), GDV, Berlin.

Guha-Sapir, D., R. Below, P. Hoyois (2009), *EM-DAT: International Disaster Database*, Université Catholique de Louvain, Brussels, *www.emdat.be.*

Güneralp, B., I. Güneralp and Y. Liu (2015), "Changing global patterns of urban exposure to flood and drought hazards", *Global Environmental Change*, Vol. 31, pp. 217–225, *http://dx.doi.org/10.1016/j.gloenvcha.2015.01.002.*

Hallegatte, S. et al. (2013), "Future flood losses in major coastal cities", *Nature Climate Change*, Vol. 3, pp. 802-806.

Hanson, S. et al. (2011), "A global ranking of port cities with high exposure to climate extremes", *Climatic Change*, Vol. 104, pp. 89–111.

Haug, O. et al. (2011), "Future building water loss projections posed by climate change", *Scandinavian Actuarial Journal*, Vol. 1, pp. 1-20.

Hirabayashi, Y. et al. (2013), "Global flood risk under climate change", *Nature Climate Change*, Vol. 3, pp. 816-821.

Hoes, O.A.C. (2007), "Aanpak wateroverlast in polders op basis van risicobeheer. Thesis (PhD), Delft University of Technology, *http://repository.tudelft.nl/assets/uuid:69de94f9–385c–4c50–beff–8cc7b955beba/ceg_hoes_20070319.pdf.*

IPCC (2014), *Climate Change 2014: Impacts, Adaptation, and Vulnerability,* the contribution of Working Group II to the Fifth Assessment Report of the Intergovernmental Panel on Climate Change, Field, C.B. et al. (eds.), Cambridge University Press, Cambridge, UK and New York.

IPCC (2012), *Managing the Risks of Extreme Events and Disasters to Advance Climate Change Adaptation: A Special Report of Working Groups I and II of the Intergovernmental Panel on Climate Change*, Cambridge University Press, Cambridge, UK, and New York, NY, USA, pp. 1-19, www.ipcc.ch/report/srex/.

Jongman, B., P. Ward and J. Aerts (2012), "Global exposure to river and coastal flooding: Long term trends and changes", *Global Environmental Change*, Vol. 22, pp. 823-835, *http://dx.doi.org/10.1016/j.gloenvcha.2012.07.004.*

Jongman, B. et al. (2014), "Increasing stress on disaster-risk finance due to large flood", *Nature Climate Change*, Vol. 4, pp. 264–268.

Keating, A. et al. (2014), *Operationalizing Resilience against Natural Disaster Risk: Opportunities, Barriers, and a Way Forward*, Zurich Flood Resilience Alliance.

Knutson, T.R. et al. (2010), "Tropical cyclones and climate change", *Nature Geoscience*, Vol 3, pp. 157–163, *www.nature.com/ngeo/journal/v3/n3/pdf/ngeo779.pdf.*

Kousky, C. and E. Michel-Kerjan (2015), "Examining flood insurance claims in the United States: six key findings", *The Journal of Risk and Insurance (forthcoming).*

Kousky, C. and L. Shabman (2015), *A Proposed Design for Community Flood Insurance*, Resources for the Future, Washington, DC, December.

Kron, W. (2015), "Global Aspects of Flood Risk Management", *Journal of Hydraulic Engineering*, Vol. 1, pp. 35-46.

Kundzewicz, Z.W. et al. (2014), "Flood risk and climate change: global and regional perspectives", *Hydrological Sciences Journal*, Vol. 59 (1), pp. 1-28.

Lloyd's (2015), *Lloyd's City Risk Index 2015-2025*, *www.lloyds.com/cityriskindex/*, accessed 10 December 2015.

Lloyd's (2012), *Lloyd's Global Underinsurance Report*, Lloyd's, London.

McGranahan, G., D. Balk, and B. Anderson (2007), "The rising tide: assessing the risks of climate change and human settlements in low elevation coastal zones", *Environment & Urbanization*, Vol. 19 (1), pp. 17–37.

Melecky, M. and C. Raddatz (2011), "How Do Governments Respond after Catastrophes? Natural-Disaster Shocks and the Fiscal Stance," *World Bank Policy Research Working Paper No. 5564*, World Bank, Washington, DC, *https://blogs.worldbank.org/files/allaboutfinance/How%20do%20Governments%20Respond%20after%20Catastrophes.pdf.*

Moncoulon, D. and M. Veysseire (2015), *Modélisation de l'impact du changement climatique sur les dommages assurés dans le cadre du régime Catastrophes Naturelles*, Caisse centrale de réassurance, Paris.

Munich Re (2015a), "Loss events worldwide 1980-2014: 10 costliest floods ordered by overall losses", *NatCatSERVICE*, Munich Re.

Munich Re (2015b), "Loss events worldwide 1980-2014: 10 costliest storms ordered by overall losses", *NatCatSERVICE*, Munich Re.

New Zealand Parliamentary Library (2014), "Canterbury in numbers: the economic effects of the earthquakes", Current issues for the 51st Parliament (September 2014), Parliamentary Library, *www.parliament.nz/resource/en-nz/00PlibC5111/df79444e096a8745ab67f8779fe0f6ff673f3840,* accessed 14 October 2015.

New York City Recovery (2015), *Community Development Block Grant Disaster Recovery: The City of New York Action Plan Incorporating Amendments 1-10 (September 23, 2015)*, *www.nyc.gov/html/cdbg/downloads/pdf/CDBG-DR_action_plan_incorporating_amendments_1-10.pdf,* accessed 13 October 2015.

OECD (2014), *Seine Basin, Île-de-France, 2014: Resilience to Major Floods*, OECD Publishing, Paris, *http://dx.doi.org/10.1787/9789264208728-en.*

Pollner, J. (2012), *Financial and Fiscal Instruments for Catastrophe Risk Management: Addressing Losses from Flood Hazards in Central Europe*, World Bank, Washington, DC.

 

RMS (2015), *Cities at Risk: A forward-looking view of wind-driven storm surge, 2010-2100 – Potential economic losses from storm surge on six key U.S. coastal cities*, Risk Management Solutions, Inc., 24 August.

RMS and Lloyd's (2008), *Coastal Communities and Climate Change: Maintaining Future Insurability*, Lloyd's, London.

Ramsbottom, D., P. Sayers and M. Panzeri (2012), "Climate Change Risk Assessment for the Floods and Coastal Erosion Sector", *Report for the UK 2012 Climate Change Risk Assessment*, *http://randd.defra.gov.uk/Document.aspx?Document=CCRAfortheFloodsandCoastalErosionSector.pdf.*

Sadoff, C.W. et al. (2015), *Securing Water, Sustaining Growth: Report of the GWP/OECD Task Force on Water Security and Sustainable Growth*, University of Oxford.

Schanz, K-U., and S. Wang (2015), *Insuring flood risk in Asia's high-growth markets: A Geneva Association research report*, The Geneva Association (The International Association for the Study of Insurance Economics), Zurich.

Schneider, F., et al. (2010), *Shadow economies all over the world: New estimates for 162 countries from 1999 to 2007*, The World Bank, Washington.

Schwarze, R. et al. (2011), "Natural Hazard Insurance in Europe: Tailored Responses to Climate Change are Needed", *Environmental Policy and Governance*, Vol. 21, pp. 14-30.

Spanger-Siegfried, E., M.F. Fitzpatrick, and K. Dahl (2014), *Encroaching tides: How sea level rise and tidal flooding threaten U.S. East and Gulf Coast communities over the next 30 years*, Union of Concerned Scientists, Cambridge, MA, *www.ucsusa.org/sites/default/files/attach/2014/10/encroaching-tides-full-report.pdf.*

Standard & Poor's Ratings Services (2015), "The Heat is On: How Climate Change Can Impact Sovereign Ratings", *Ratings Direct*, 25 November.

Surminski, S. and J. Eldridge (2014), "Flood insurance in England – an assessment of the current and newly proposed insurance scheme in the context of rising flood risk", *Centre for Climate Change Economics and Policy Working Paper No. 161/Grantham Research Institute on Climate Change and the Environment Working Paper No. 144*, Centre for Climate Change Economics and Policy, Leeds.

Swiss Re (2015), "Natural catastrophes and man-made disasters in 2014: convective and winter storms generate most losses", *Swiss Re sigma No 2/2015*, Swiss Reinsurance Company Ltd., Zurich, *http://media.swissre.com/documents/sigma2_2015_en_final.pdf.*

Swiss Re (2012), *Flood – an underestimated risk: Inspect, Inform, Insure*, Swiss Reinsurance Company Ltd., Zurich.

Swiss Re (2009), *The effects of climate change: An increase in coastal flood damage in Northern Europe*, Swiss Reinsurance Company Ltd., Zurich, *http://media.swissre.com/documents/the_effects_of_climate_change_an_increase_in_coastal_flood_damage_in_northern_europe.pdf.*

Tebaldi, C., B. Strauss and C. Zervas (2012), "Modelling sea level rise impacts on storm surges along US coasts", *Environmental Research Letters*, Vol. 7 (1), IOP Publishing Ltd.

Te Linde, A.H. et al. (2011), "Future flood risk estimates along the river Rhine", *Natural Hazards Earth System Sciences*, Vol. 11, pp. 459–473.

Thakkar, K. (2015), "Chennai floods: Carmakers like Ford and BMW struggle to restart work after screeching halt", *The Economic Times*, 15 December, *http://articles.economictimes.indiatimes.com/2015-12-15/auto/69062062_1_production-loss-bmw-india-ford-india*, accessed 22 March 2016.

Thieken, A.H. et al (2016), "The flood of June 2013 in Germany: how much do we know about its impacts?", *Natural Hazards and Earth Systems Science*, Discuss., doi:10.5194/nhess-2015-324, in review, 2016.

Turnham, J. et al. (2011), *Housing recovery on the Gulf Coast – Phase II: Results of property owner survey in Louisiana, Mississippi and Texas*, Department of Housing and Urban Development, Washington, DC, August.

Von Peter, G., S. von Dahlen and S. Saxena (2012), "Unmitigated disasters? New evidence on the macroeconomic cost of natural catastrophes", *BIS Working Papers No. 394*, Bank for International Settlements, Basel, *www.bis.org/publ/work394.pdf.*

Wilby, R. and R. Keenan (2012), "Adapting to flood risk under climate change", *Progress in Physical Geography* , Vol. 36 (3), pp. 348–378, Sage.

Wilson, S.G., and T.R. Fischetti (2010), *Coastline population trends in the United States: 1960 to 2008*, US Census Bureau, Washington, DC, May, *www.census.gov/prod/2010pubs/p25-1139.pdf.*

World Bank (2016), *World Bank FY2016 country and lending groups*, *http://data.worldbank.org/about/country-and-lending-groups,* accessed 16 March 2016.

World Bank and Thai Ministry of Finance (2012), *Rapid Assessment for Resilient Recovery and Reconstruction Planning*, World Bank, Bangkok.

# *Chapter 3*

# Insuring flood risk

*This chapter provides an overview of insurance arrangements for covering flood risk across countries, including the role of private insurance markets and governments in providing coverage, the form of insurance coverage available, and the level of coverage. It identifies the significant "financial protection" gap that exists for flood risk and outlines the factors that make flood risk a particularly difficult peril to cover.*

---

The statistical data for Israel are supplied by and under the responsibility of the relevant Israeli authorities. The use of such data by the OECD is without prejudice to the status of the Golan Heights, East Jerusalem and Israeli settlements in the West Bank under the terms of international law.

There is a wide variety of approaches across countries to protecting households and businesses against flood risk. In many countries, private insurance companies offer coverage for flood-related damages and losses, either as part of standard property and business interruption policies, or available as an optional add-on to such policies. In some countries, coverage for flood damage may only be available from a public insurer, especially for properties deemed to be at high-risk of flooding. In other countries, government assistance may be the only source of compensation available for losses from flood events. In countries with lower levels of insurance market development and penetration, micro-insurance might play an important role in providing financial protection.

These different approaches to financial protection have been designed with the aim of achieving different policy objectives, such as broad availability and affordability of coverage, solidarity in terms of loss-sharing across regions, establishment of clear incentives for risk reduction and/or significant transfer of risk to private markets. There are clear trade-offs between these different approaches. For example, broad availability and affordability of coverage and/or solidarity across regions usually entails some form of cross-subsidisation across policyholders with implications for the strength of incentives to encourage risk reduction. In some instances, a reliance on private markets (and full risk-based pricing) may come at the expense of the availability of affordable coverage for high-risk properties.

## 3.1 Financial protection against flood risk across countries

### *Private insurance coverage for flood risk*

The insurance coverage of flood risk in a number of OECD and non-OECD countries is exclusively or primarily provided by private insurance companies (see Table 3.1). In some countries, there may be some differentiation in terms of the type of flood perils covered (e.g. inland vs. coastal flooding, overland vs. sewer back-up, etc.). Some insurance policies may include coverage for additional living expenses in cases where the level of damage to residential structures impedes access to the property, either as an additional option or part of standard coverage. Coverage for flood damage is also available for motor vehicles in most countries, either as part of standard coverage or as an optional add-on. For businesses, coverage of property and contents for flood risk and for business interruption is the most common form of financial protection of flood risk.

Standard residential property insurance policies in a few countries (Australia (flash flooding), Austria (basic amount), Belgium, Denmark, Finland, France, Iceland, Israel, Latvia, New Zealand, Norway, Poland, Russia, Spain, Switzerland and the United Kingdom) are automatically extended to cover flood risk, usually bundled as coverage for all or most natural perils. In Japan and Turkey, flood coverage is usually included in standard residential property policies in practice (although bundling is not a formal requirement for insurers). In Switzerland, insurance coverage for residential and commercial buildings against a number of natural perils, including flood risk, is mandatory in 22 of 26 cantons (coverage for contents and motor vehicles is not mandatory but nevertheless widely used). In Belgium, insurers may not extend coverage to high-risk properties built after the completion of risk maps (Wharton Risk Management and Decision Processes Center, 2016).

Table 3.1. **Insurance arrangements for flood risk**

| Country | Coverage Provider | | Form of Coverage | | Description |
|---|---|---|---|---|---|
| | Private | Public | Automatic extension | Optional Add-On | |
|  *indicates motor vehicle coverage;*  *indicates residential property coverage; and*  *indicates commercial property coverage.* | | | | | |
| Australia |  | |  |  | A distinction is made between damage caused by flash floods and riverine floods. Coverage for flash flooding is generally included in standard policies. Insurers are required to offer riverine flood coverage as part of standard cover, but may derogate if disclosed to the policyholder. Consumers have a wide choice of products: 86% of policies selected by consumers have flood cover as a standard inclusion, with no opt-out option; 7% of policies are sold with flood cover as an inclusion, but with the opportunity to opt-out; and 7% of policies sold derogate flood cover entirely. Flood insurance is sold on a risk-based pricing basis and is not cross subsidised. |
| Austria |  | |  |  | "First risk" coverage for flood damage is automatically extended to standard residential fire policies. Extended coverage is available on an optional basis. |
| Belgium |  | |  | | Insurance companies may not extend standard policies to high-risk properties. |
| Canada |  |  | |  | The availability of private coverage for flash and riverine flooding for residential property is new and not yet available for all properties (coverage for coastal flood is not available). Motor vehicle insurance, including coverage for flood risk, is provided by public insurers in some provinces. |
| Chile |  | | |  | Insurance for flood is bundled with other natural disaster perils and available as an optional add-on to standard coverage. |
| Costa Rica |  |  | |  | Most flood insurance coverage is provided by a public insurer (operating similar to a private insurer) that provides various types of insurance coverage. The public insurer will not provide coverage for high-risk structures (e.g. located too close to coasts or rivers) |
| Czech Republic |  | | |  | Insurance companies are unwilling to offer flood coverage in flood-prone areas. |
| Denmark |  |  |  | | A mandatory charge is attached to all fire policies and used to provide compensation for damage from storm surge and inland flooding through the Danish Storm Council where private insurers do not provide coverage. |
| Estonia |  | | |  | There is one region where flood insurance availability may be limited due to regular flooding. |
| Finland |  | |  | | Since 2014, flood insurance has been included in standard residential property coverage although coverage is only provided for damage above a certain threshold. |
| France |  |  |  | | Private insurers automatically extend coverage to include natural disasters (at a flat rate) and can reinsure up to 50% of their natural disaster exposure with a public reinsurer (CCR). |

| Country | Coverage Provider | | Form of Coverage | | Description |
|---|---|---|---|---|---|
| | Private | Public | Automatic extension | Optional Add-On | |
| *[vehicle] indicates motor vehicle coverage; [residential] indicates residential property coverage; and [commercial] indicates commercial property coverage.* | | | | | |
| Germany | [vehicle] [residential] [commercial] | | | [vehicle] [residential] [commercial] | Standard policies exclude storm surge and flash flooding. Coverage for a set of natural perils is bundled and made available as an optional add-on. |
| Hungary | [vehicle] [residential] [commercial] | [residential] | | [vehicle] [residential] [commercial] | Insurance companies are unwilling to offer flood coverage in flood-prone areas. A public financial protection fund has been established to provide insurance for high-risk residential properties. |
| Iceland | [vehicle] | [residential] [commercial] | [residential] [commercial] | [vehicle] | ICI provides coverage for natural disaster risks, including floods, as an automatic extension to all residential and commercial property insurance policies. Coverage against fire (and therefore flood and other natural perils) is mandatory for commercial and residential property. |
| Ireland | [vehicle] [residential] [commercial] | | | [vehicle] [residential] [commercial] | Insurance companies are unwilling to offer flood coverage in flood-prone areas. |
| Israel | [vehicle] [residential] [commercial] | [vehicle] [residential] [commercial] | [vehicle] [residential] | [commercial] | Flood insurance is part of broader natural risks coverage. |
| Italy | [vehicle] [residential] [commercial] | | | [vehicle] [residential] [commercial] | Coverage for natural disaster risks is available as an optional extension to standard policies. |
| Japan | [vehicle] [residential] [commercial] | | | [vehicle] [residential] [commercial] | While coverage for flood damage is optional, most standard fire policies include coverage for flood damage under a single premium. |
| Latvia | [vehicle] [residential] [commercial] | [residential] | [vehicle] [residential] [commercial] | | Insurance companies are unwilling to offer flood coverage in flood-prone areas or only with high deductibles. |
| Mexico | [vehicle] [residential] [commercial] | | | [vehicle] [residential] [commercial] | Flood coverage is bundled with coverage for other hydro-meteorological risks (e.g. hurricanes) as an optional add-on to standard property policies. |
| Netherlands | [vehicle] [residential] [commercial] | | | [residential] | The availability of private flood insurance for residential property is new and limited. |
| New Zealand | [vehicle] [residential] [commercial] | [residential] | [vehicle] [residential] [commercial] | | The public Earthquake Commission provides coverage for damage to land in and near residential properties and access ways. Private insurers provide coverage for flood damage to structures. |
| Norway | [vehicle] [residential] [commercial] | | [vehicle] [residential] [commercial] | | The Norwegian Natural Perils Pool has been established to pool natural disaster losses among private insurers. |
| Peru | [vehicle] [residential] [commercial] | [residential] | | [vehicle] [residential] [commercial] | |
| Philippines | [vehicle] [residential] [commercial] | [vehicle] [residential] [commercial] | | [vehicle] [residential] [commercial] | Some general insurance coverage is provided by publicly-owned insurance companies. These companies are the only providers of flood insurance for high-risk properties. |
| Poland | [vehicle] [residential] [commercial] | | [vehicle] [residential] [commercial] | | |
| Portugal | [vehicle] [residential] [commercial] | | | [vehicle] [residential] [commercial] | Insurance companies may not offer flood coverage or may stipulate unaffordable tariffs. |
| Russia | [vehicle] [residential] [commercial] | | [vehicle] [residential] [commercial] | | Insurance is not available for structures in flood zones (where in violation of construction permits). Insurance companies may exclude flood coverage in flood-prone |

| Country | Coverage Provider | | Form of Coverage | | Description |
|---|---|---|---|---|---|
| | Private | Public | Automatic extension | Optional Add-On | |
| [Motor] indicates motor vehicle coverage; [Residential] indicates residential property coverage; and [Commercial] indicates commercial property coverage. | | | | | |
| | | | | | areas. |
| Spain | Motor, Residential, Commercial | Motor, Residential, Commercial | Motor, Residential, Commercial | | An extraordinary risk cover clause is mandatorily included in property, life and personal accident policies, and a mandatory surcharge is applied. The risk is assumed by CCS (provided it is not assumed by the company on its own). CCS is provided with an unlimited state guarantee in case its resources are exhausted (never used). |
| Switzerland | Motor, Residential, Commercial | Motor, Residential, Commercial | Motor, Residential, Commercial | | Coverage for natural disaster losses is mandatory for residential and commercial buildings in 22 of 26 cantons. Insurance coverage for buildings is provided by private insurers in 7 cantons. In the other 19 cantons, natural disaster insurance coverage for buildings is provided by canton monopole insurers only. Coverage of contents is provided by private insurers in all cantons. Contents and motor vehicle insurance coverage is not mandatory. |
| Turkey | Motor, Residential, Commercial | | | Motor, Residential, Commercial | The coverage of flood risk is not required as part of standard fire policies although, in practice, most policies are automatically extended to cover flood risk. However, insurance companies may not offer flood coverage in flood-prone areas or may stipulate extra conditions. |
| United Kingdom | Motor, Residential, Commercial | Residential | Motor, Residential, Commercial | | Private insurers can transfer risks related to their coverage of certain high-risk properties to Flood Re, an industry established pool. |
| United States | Motor, Residential, Commercial | Residential | Motor | Residential, Commercial | The National Flood Insurance Program (NFIP) provides flood coverage for residential properties in eligible communities. Private insurers may also provide alternative coverage or excess coverage for amounts above the maximum level of NFIP coverage (excess flood insurance) as well as coverage for additional living expenses (which is not covered by the NFIP). |
| Viet Nam | Motor | | | Motor | Flood insurance coverage is generally only available for motor vehicles. |

*Source:* Most of the information was taken from country responses to an OECD questionnaire on the financial management of flood risk (2015). Additional information was also taken from UNISDR (2015), Wharton Risk Management and Decision Process Centre (2016), Maccaferri, S., J. Carboni and F. Campolongo (2012), Insurance Bureau of Canada (2015) and Swiss Re (2015b).

In other countries, insurance protection against flood risk is offered as an optional add-on to standard property policies, either as a single peril or in combination with other disaster risks. In Germany, the optional add-on is for insurance coverage against all natural catastrophes (*Elementarschadenversicherung*). Similarly, optional coverage for a set of natural catastrophes is available in the Philippines. In Austria, a basic amount of coverage for flood risk is automatically included on a first-loss basis with the option available to purchase additional coverage. In Austria, Costa Rica, Czech Republic, Ireland, Portugal, Turkey (amongst other countries), optional coverage for properties in flood-prone areas is not always available or available only with high deductibles, at high-cost and/or upon the implementation of specific risk prevention measures. In Canada and the Netherlands, flood insurance for residential properties has only recently become available (previously, the only compensation available was provided by the public sector).

In a number of countries (Czech Republic, Portugal, United Kingdom, United States), lender practice or legislation requires some or all properties with mortgages to be protected against flood risk. In the Czech Republic, Ireland, New Zealand, Portugal, Sweden and the United Kingdom, mortgage lenders always (or generally) require borrowers to obtain insurance protection against flood risk. In the United States, federally-regulated mortgage lenders are legally required to ensure that borrowers with properties in flood-prone areas are protected by flood insurance.[1]

In almost all countries, premiums charged by private insurers vary (to some extent) with the level of risk, although with varying levels of granularity. In Switzerland, premiums charged by private insurers in the 7 cantons without a cantonal monopole insurer are established by regulation at a flat rate (although that level is established based on an assessment of overall exposure across a number of perils including flood). In other countries, limits to the capacity of private insurers to assess flood exposures and/or the practice of bundling coverage with other perils limits the alignment of premiums to the level of risk.

### *Public insurance of flood risk*

In a number of countries, the public sector provides financial backing for the insurance coverage of flood risk, either as a direct insurer or reinsurer for all or a sub-set of properties (see Table 3.1). Iceland and 19 of 26 Swiss cantons offer bundled direct insurance underwritten by a public entity for all or most natural perils for all residential and commercial properties. In France, reinsurance for all natural disaster risks is offered by the public *Caisse centrale de reassurance* (CCR) for up to 50% of the losses, although private insurance companies are not required to purchase reinsurance from CCR. Similarly in Spain, the *Consorcio de Compensación de Seguros* (CCS) manages the "extraordinary risks" insurance coverage which offers direct insurance for flood and other extraordinary risks by means of coverage that is mandatorily included in property, life and personal accident policies issued by private companies. Private insurers may choose to retain the extraordinary risks or transfer the risk to CCS (however, should they decide to transfer these risks to CCS, they must transfer all extraordinary risks). In all these cases, insurance coverage is either mandatory or provided as an automatic extension to property, business interruption or motor vehicle policies. Premiums charged to households and businesses are generally flat (i.e. do not vary with the level of risk), although with some level of variation in the case of Switzerland according to the exposure. In France, premiums are flat although deductibles are increased for repeated claims due to the same peril in communes that do not have a *plan de prévention des risques* (risk prevention plan) (Fédération française des sociétés d'assurances, 2016).

In Korea, a public scheme (operated by a private insurance company) provides coverage for storm and flood risk to residential properties. New Zealand's Earthquake Commission provides direct insurance coverage for damage to residential land from flooding (along with coverage for residential land and structures against earthquake and several other risks). In Thailand, the National Catastrophe Insurance Fund reinsures a portion of risks covered in catastrophe insurance policies offered by the private sector for flood, earthquake and windstorm damage. In the United States, direct flood insurance is offered through the public National Flood Insurance Program (NFIP). Premiums are generally risk-based, although with various exceptions (see Box 3.1).

**Box 3.1. US National Flood Insurance Program premiums**

The National Flood Insurance Program (NFIP) was established in 1968 to provide flood insurance coverage for residential and commercial properties in floodplains in response to the withdrawal of such coverage by private insurers. The insurance is offered only in communities that agree to a set of flood management conditions, including building standards and floodplain management standards approved by the Federal Emergency Management Agency (FEMA). As of 31 July 2015, the NFIP had over 5 million insurance policies in force providing almost USD 1.3 trillion in insurance coverage.

The majority of policyholders pay premiums based on the level of risks (approximately 80% of all policyholders) (National Academies of Sciences, Engineering, and Medicine, 2015)). These premiums are based on flood insurance rate maps (FIRMs) produced (and updated) by FEMA. Risk-based premiums are charged for structures built after the completion of the relevant FIRM for their community (FIRM's are developed and updated over time), based on the location of the insured property within the floodplain. Structures constructed after the completion of a FIRM must meet FEMA building standards in order to access insurance under the program.

Structures built before the completion of a community's FIRM ("pre-FIRM properties") generally do not meet FEMA building standards and/or were constructed without considering the base flood elevation for a 1-in-100 year flood. These properties are therefore much less protected against flood. Full risk-based premiums for such properties would be extremely expensive and therefore pre-FIRM property owners benefit from subsidised premiums - although these rates are still generally higher than rates charged on properties built after the completion of a FIRM and to FEMA building standards (the average annual subsidised premium for pre-FIRM properties was approximately USD 1 224, while the average annual premium for post-FIRM properties paying full-risk rates was approximately USD 492 (GAO, 2014)). The pre-FIRM properties with subsidised rates have accounted for a significant portion of losses over the history of the NFIP (Michel-Kerjan, 2010).

In July 2012, the *Biggert-Waters Flood Insurance Reform Act of 2012*, requiring the NFIP to immediately charge full risk-based premiums on all policies, was signed into law. However, some elements of the rate increase was repealed in March 2014 under the *Homeowner Flood Insurance Affordability Act of 2014*, which lowers the rate increases for some policies and prevents some future rate increases (see Box 4.9).

*Source:* National Academies of Sciences, Engineering, and Medicine, 2015; GAO, 2014; Michel-Kerjan, 2010.

In Hungary and the United Kingdom, public support for flood insurance is only offered for high-risk properties as coverage for lower risk properties is available from the private sector. In Hungary, *Wesselenyi Miklos Ar-es Belvizvedelmi Alap* provides flood insurance for residential properties in high-risk areas. In the United Kingdom, a not-for-profit reinsurance pool, Flood Re, has been established by industry to cover flood damage to high-risk residential properties (see Box 3.2). While not a public entity, Flood Re is formalised through legislation.

Box 3.2. **Coverage provided by UK Flood Re**

The insurance sector in the United Kingdom has established a flood reinsurance fund (Flood Re) to provide affordable reinsurance cover for high-risk residential properties. Flood insurance is included as an automatic extension to home insurance policies covering fire, theft, etc. Flood Re provides a reinsurance option that insurers can access to cover their flood exposure related to the residential properties that they insure. Insurers have the option to transfer the premiums (and claims liability) from eligible policies to Flood Re or retain the risk themselves. Flood Re was launched in April 2016 and will operate until 2039.

Premiums for the reinsurance coverage provided by Flood Re are set at rates that vary with the value of the property (rather than risk-level) in order to ensure that premium subsidies are targeted to lower-income households. Therefore, insurance companies seeking reinsurance from Flood Re for two properties of similar value (i.e. are part of the same Council Tax band) would pay the same amount for the reinsurance cover, even if the households face very different levels of flood risk.

Flood Re is funded by the premiums collected from insurers on reinsured policies and a levy collected from all insurers over five years based on the insurer's market share. The levy has been established based on an estimate of the existing cross-subsidy for high-risk residential properties previously included in all home insurance policies, with the aim of ensuring that the levy does not lead to a general increase in home insurance premiums. The funds are used to purchase reinsurance coverage on international markets. The industry levy and the premiums will be reviewed every five years with the aim of ensuring that Flood Re is adequately funded and that Flood Re is transitioning towards risk-reflective pricing, consistent with the longer-term objective of returning to a free market for flood insurance.

*Source*: Flood Re (2015).

Similar pools have been established by insurers (and formalised in legislation) to payout claims related to natural disaster losses in Norway (Norwegian Natural Perils Pool) and Romania (Insurance Pool against Natural Disasters). However, unlike the United Kingdom, these pools cover a broader set of natural perils and all properties (not just high-risk properties). In Belgium, private insurers provide coverage for natural disasters, although the government provides a guarantee to private insurers to cover losses for extreme events above a specific threshold (up to EUR 280 million per insurer and event if damage per insurer and event exceeds EUR 3 million plus 0.35 times the premium income of the insurer (Schwarze et al., 2011)).

### *Microinsurance*

In many developing countries, insurance coverage for residential property and contents is generally not available or is only available at a cost above the willingness-to-pay (i.e. the maximum amount an individual is willing to pay for financial protection) of significant portions of the population. In these countries, microinsurance may provide a mechanism for offering some financial protection against flood risk. Such products can potentially be offered at an affordable price where payouts are relatively small and calculated based on parametric weather triggers (index insurance) rather than indemnity triggers and where efficient distribution channels are available. However, few products have thus far been able to demonstrate economic viability and/or generate significant scale and many microinsurance initiatives have been dependent on continued support from donor funding.

Microinsurance providing financial protection against multiple disaster perils is available in some countries. In India, *Afat Vimo* provides protection of up to INR 95 000 against building and contents damage, stock-in-trade, personal accident and death from various natural disaster risks, including floods. In the aftermath of Cyclone Phailin in 2013, *Afat Vimo* settled 125 claims and paid out INR 400 507 to individuals impacted by the event (Gupta and Agrawal, 2015). In the Philippines, which has the highest level of microinsurance penetration in Asia, microinsurance providers played a significant role in providing financial protection to those affected by Typhoon Haiyan (Yolanda) in 2013. Providers of calamity microinsurance coverage paid out over PHP 453 million to close to 110 000 policyholders at an average of just over PHP 4 000 per policyholder (Swiderek and Wipf, 2015). In Indonesia (Jakarta), a parametric trigger-based microinsurance product (*Asuransi Wahana Tata*) was developed to specifically provide financial protection against flood risk for vulnerable populations (Malagardis, 2015) although the product was eventually discontinued as the costs of providing coverage were beyond individuals' willingness-to-pay (Lamond and Penning-Roswell, 2014).

In Bangladesh, index-based flood insurance coverage (based on water depth and flooding duration) is provided to a local community organisation on behalf of local households, allowing for a more simplified structure for providing compensation (Swiss Re, 2015a). In Haiti, MiCRO provides index-based coverage to a microfinance institution (*Fonkoze*) and its borrowers against earthquake, wind and excess rainfall (Guy Carpenter, 2015). In Peru, an innovative approach that allows insurance payments to be made before the occurrence of a disaster has been developed. The Extreme El Niño Insurance Product, offered by *La Positiva*, a private insurance company in Peru, provides payouts that are triggered by the severe increases in sea surface temperatures that usually occur during an El Niño year and generally result in heavy precipitation and flooding. By paying out in advance of flooding, the funding allows policyholders to finance risk reduction measures to protect themselves against the potential losses associated with extreme El Niño years (OECD, 2015).

### *Public compensation*

In most countries, public compensation and/or financial assistance is provided to households and businesses to mitigate the financial impact of flood events, particularly major events (national governments also provide compensation and financial assistance to sub-national governments in many countries – which is discussed in Chapter 5).

In countries where insurance coverage for flood risk is generally unavailable, government compensation absorbs the vast majority of private losses from flood events. In the Netherlands, where flood damages have generally been considered uninsurable, the national government provides partial compensation for flood damages. The Calamities and Compensation Act (*Wet Tegmoetkoming Schade bij Rampen en Zware Ongevallen - WTS*) allows the government to provide compensation to those impacted by freshwater flood events (compensation for saltwater flood damages is excluded from the WTS, although other compensation may be made available). The government decides on the amount of compensation available to affected households for a given event, up to the legislated aggregate of EUR 450 million. In Canada, provincial and territorial governments provide compensation and financial assistance to households that have suffered losses (generally only when the losses are uninsured). As damage from overland flooding is excluded from most residential insurance policies across Canada (coverage for sewer back-up is available in most provinces), flood losses are usually uninsured and therefore eligible for compensation under such programs.

Government compensation or financial assistance is also often provided in countries where flood insurance is available, either through pre-determined programs and funding mechanisms or on an *ad hoc* basis. In Australia, compensation and financial assistance is often available from state and territorial governments. In addition, for significant events (as defined by the national government), two payments by the national government are available to support recovery (Australian Government Disaster Recovery Payment) and lost income (Disaster Recovery Allowance). The United States federal government offers similar assistance to those that face unemployment as a result of a Presidentially-declared disaster (Disaster Unemployment Assistance). The United States also offers federal loans through the US Small Business Administration to homeowners (up to USD 200 000) and businesses (up to USD 2 million) for repair or replacement of damaged buildings. These loans are available once to all households and businesses affected by flood damages, although subsequent loans are only available if the homeowner or business has secured flood insurance coverage (GAO, 2014). The loans are provided for extended tenures with low interest rates available to those unable to otherwise secure credit (Kousky, Michel-Kerjan and Raschky, 2014).

In Austria, state governments may provide compensation for flood losses to private property, a share of which can be reimbursed to state governments by the federal Austrian Catastrophes Fund. Approximately 4% of payments from the Austrian Catastrophes Fund have been used for the compensation of private losses (with the remaining 96% spent on public losses and prevention). In Belgium, the government may provide compensation and/or financial assistance to individuals affected by floods through the *Caisse nationale des Calamités* (natural disaster fund) if the compensation provided through private insurance arrangements is deemed insufficient. In Germany, there are no formal legislated requirements to provide compensation, although governments have provided compensation and financial assistance to households for damage from past major flood events. Surminski et al. (2014) found that 34 *ad hoc* compensation schemes in European Union countries for flood damage were notified to the European Commission since 2007 with a total value of EUR 1.7 billion in compensation provided.[2]

## 3.2 Underinsurance of flood risk

While property insurance companies, governments and micro-insurance providers offer insurance coverage against flood losses in most OECD and many other countries, significant gaps remain in terms of the share of flood losses that are covered by insurance. This is also true for many other natural disasters although there is some evidence that the gap is particularly significant for flood (as well as earthquake) losses.

While flood losses (not including losses related to storm surge, which are considered separately in statistics on disaster losses) accounted for approximately 19% of total disaster losses between 2005 and 2018, flood losses accounted for close to 23% of all uninsured losses suggesting that flood losses are less insured than other losses (see Figure 3.1).[3]

Figure 3.1. **Total and uninsured losses by disaster type**

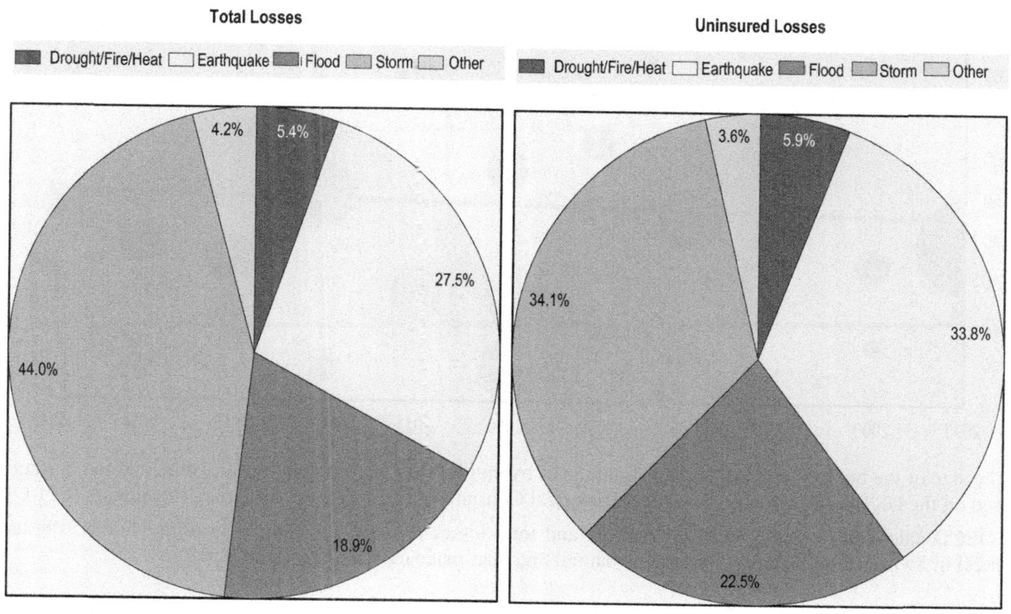


*Notes:* It should be noted that the figures take account of losses insured by the National Flood Insurance Program (i.e. the share of insured losses includes losses insured by the NFIP).

*Source*: OECD calculations based on insured losses and total damages reported for natural disasters (floods, storms, earthquake, droughts/fires/heat waves and other natural disasters) in Swiss Re sigma annual reports on natural and man-made catastrophes (2005-2015).

However, unlike in the case of earthquake losses (and losses from other disasters), there is also some evidence that the share of flood losses that are insured has actually declined in the past decade (although the share of losses insured can vary significantly from year-to-year). Just over 21% of losses related to flooding between 2005 and 2009 were insured relative to approximately 15% between 2010 and 2014 (see Figure 3.2).

The level of insurance penetration varies substantially across countries. In countries where flood insurance is provided as an optional add-on to residential property insurance policies, take-up rates are generally very low. For example, market penetration for the natural disaster insurance add-on in Germany is estimated at 38% despite being available to households in more than 99% of the country (GDV, 2015b). Market penetration for flood insurance coverage is also relatively low in Turkey and Austria. Estimates for other European countries, including Bulgaria, Greece, Italy and Luxemburg also found low penetration.

Requirements for flood insurance coverage attached to mortgages have led to broad coverage of flood risk in Ireland and Sweden where penetration rates are above 90% (Maccaferri, Carboni and Campolongo, 2012). However, in other countries with mortgage-related requirements for flood insurance, such as the Czech Republic and Portugal, penetration rates remain relatively low (and would likely be even lower without the mortgage-related requirements). In the United States, it is estimated that approximately 50% of all residential properties in the Special Flood Hazard Area (SFHA) are covered by flood insurance (where mortgage requirements are in place) while less than 1% of homes in the 500-year flood zone are covered (Bin and Landry, 2013).

Figure 3.2. **Trends in the share of losses that are insured by disaster type**

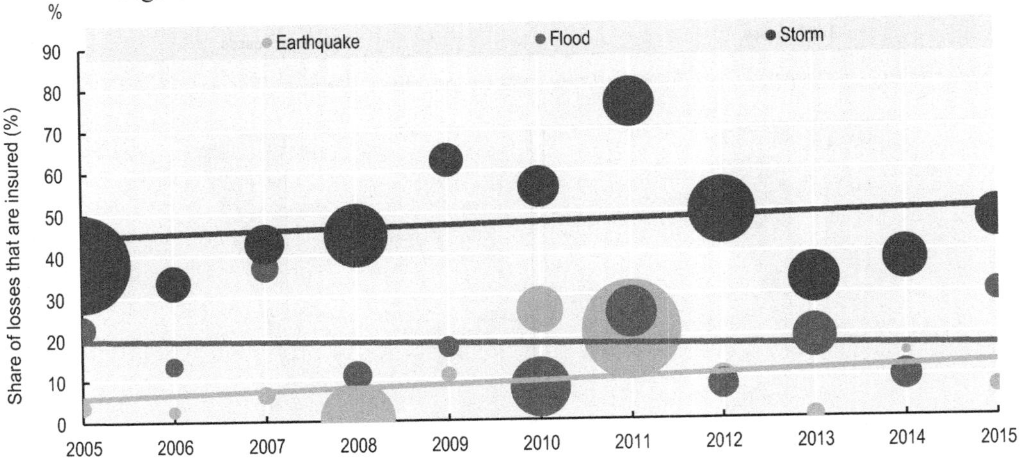


*Notes:* The size of the bubbles represents the magnitude of overall losses reported in that year, converted to constant 2015 USD based on the US Bureau of Labor Statistics' Historical Consumer Price Index for All Urban Consumers (CPI-U).

*Source*: OECD calculations based on insured losses and total losses reported for natural disasters (floods, storms, and earthquakes) in Swiss Re sigma annual reports on natural and man-made catastrophes (2005-2015).

In countries where flood risk are automatically included in standard building and contents insurance for households and businesses, penetration rates are generally higher. In the United Kingdom, take-up rates for residential property insurance are over 90% while penetration of insurance for home contents (not required by mortgage lenders), ranges from 44% to 90% (DEFRA, 2013). Penetration rates in other countries where flood risk are included in standard coverage, including Israel and Latvia, are also relatively high (see Figure 3.3).

Figure 3.3. **Estimates of residential flood insurance penetration (by form of offering)**

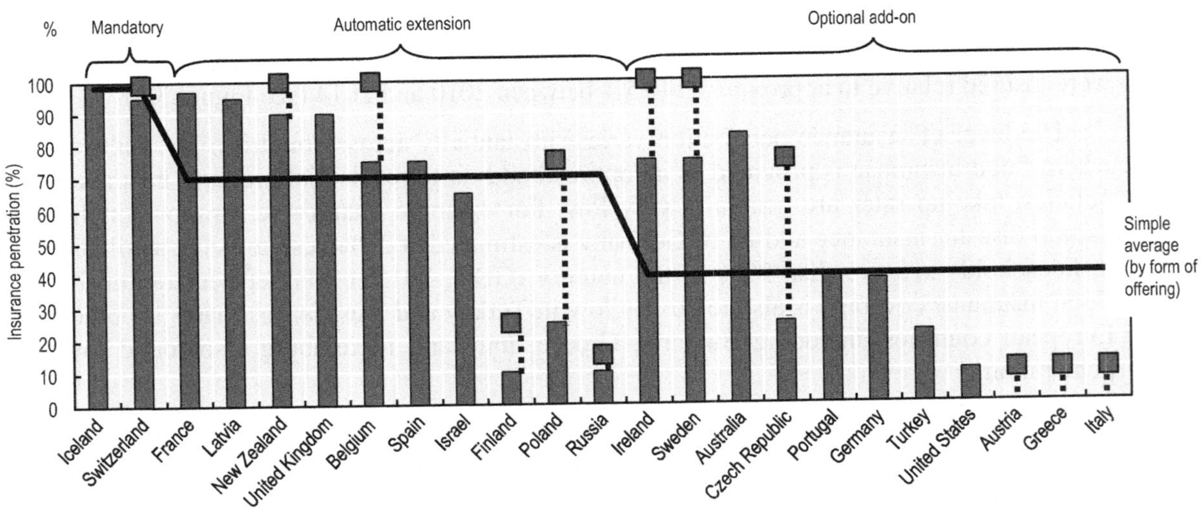


*Notes:* The dashed lines represent the range of estimated penetration rates (minimum estimate to maximum estimate). The solid line indicates the simple average across each type of offering.

*Source*: Most of the information on penetration rates (share of households with flood insurance coverage) was taken from country responses to an OECD questionnaire on the financial management of flood risk (2015). The estimated penetration rate for Australia is from OECD (2015); the estimated ranges for penetration rates in Belgium, Czech Republic, Finland, Greece, Ireland, Italy and Poland are from Maccaferri, Carboni and Campolongo (2012); for the UK, the estimate is from DEFRA (2013); and for Germany, the estimate is from GDV (2015b).

In developing countries, flood insurance penetration rates are even lower as a result of lower levels of insurance market development (as well as more limited capacity to pay for many households). For example, in East Asia and the Pacific, the average non-life insurance penetration rate (i.e. premiums as a share of GDP for broad non-life coverage, not just floods) is approximately 50% of the penetration rate in Europe and 35% of the penetration rate in North America (Jha and Stanton-Geddes, 2013). In Latin America, non-life penetration rates range from less than one-quarter of the United States' penetration rate in Peru and the Dominican Republic to approximately half of the United States' rate in Brazil, Colombia and Chile (Swiss Re, 2016).

## 3.3 Challenges to insuring flood risk

The offering of insurance coverage for a given risk is usually economically viable only where certain criteria (or "principles of insurability") are generally met (Swiss Re, 2012; Insurance Europe, 2012). These criteria include:

- Risks must be ***quantifiable***: the probability of occurrence of a given peril, its severity and its impact in terms of damage and losses, given the structural characteristics and vulnerabilities of the insured assets, must be assessable.
- A sufficiently large community with assets at risk can be established to share the risk (***mutuality***), allowing for sufficient diversification of the risk based on differences across the community in terms of risk exposure.
- Risks must occur ***randomly***: the time and location of an insured event must be unpredictable and the occurrence must be independent of the will of the insured.

The extent to which the characteristics of a given risk exposure meets these criteria (among other factors) will impact whether insurance companies can collect the amount of premiums necessary to cover the total losses of a community of insureds (along with administrative costs and returns to investors, where provided by private insurance companies). In other words, the actuarially-sound premium rates charged to policyholders must be both within their willingness-to-pay for protection and provide sufficient funds in aggregate to cover losses and other costs.

Catastrophe risks do not always meet these criteria as a number of factors lead insurance companies to charge premiums for disaster insurance coverage that is beyond the willingness-to-pay for such coverage. Among disaster risks, floods pose particular challenges in terms of insurability for a number of reasons. According to Swiss Re (2012), "no other peril defies the basic principles of insurability to the same degree."

The following sections will outline: i) factors that drive up the price for flood insurance coverage; and ii) factors that lower the willingness-to-pay of consumers. This mismatch between demand and supply has played a role in limiting the availability of flood insurance from the private sector in a number of countries that face significant potential losses from floods, including Canada, the Netherlands and the United States (GAO, 2014; Siefert et al., 2013).

### *Factors affecting the price of flood insurance*

There are a number of factors that affect the price at which insurance companies are willing to offer coverage for a given risk, including the size of expected losses (economic viability), the diversity of the pool of risks covered (mutuality and randomness) as well as the level of uncertainty in estimating expected losses (quantifiability).

*Size of expected losses*

The expected loss on an insurance policy providing coverage for a given property depends on the frequency of damaging events as well as the extent of possible damage caused by such events. Flood-prone areas are generally (particularly outside major urban areas in developed countries) not protected for events beyond a 1-in-100 year return period which is a relatively high level of frequency for an insurance loss (by comparison, the average return period for fire is 1-in-340 years (Green and Penning-Roswell, 2004)). In the United States, a property located in a flood zone has a 26% chance of being flooded over the life of a 30-year mortgage, compared with a 1% chance of suffering a fire loss (Collins and Simpson, 2007). The average claim size for flood also tends to be larger than other natural disasters. In Australia, the average claim during the Queensland inland flooding was AUD 45 374 compared to AUD 15 959 after Cyclone Yasi (wind and flood) and approximately AUD 6 000 – 8 000 for major hail and other storms (Allianz Australia Insurance Ltd., 2011).

The frequent return periods of flooding in high-risk areas and the large level of potential damage related to each flood event means that risk-based premiums become very high for high-risk properties. The actuarially-sound annual premiums for properties prone to severe flooding (more frequently than 1-in-50 years) or that might be destroyed by storm surge more than once in a hundred years would exceed 1% of the value of the property. Even if expected damage from a 1-in-50 year flood is only 25% of the value of a property, expected annual losses on that policy would still be equivalent to 0.5% of the value of the home (or 2 500 for a home with a value of 500 000).

As noted above, climate change is expected to increase both the frequency and severity of flooding which will translate into higher expected losses. A number of studies have translated the potential increase in expected losses into estimates of the resulting increase in premiums for specific risks in hazard-prone regions of the world:

- In the Netherlands, hypothetical risk-based premiums of approximately EUR 34 per year (on average) would need to increase by 93-102% in 2040 and 641-797% in 2100 (depending on demographic assumptions) in some areas in response to a 2-3 times increase in estimated flood probability between 2015 and 2040 and a 16-20 times increase in estimated flood probability by 2100 (based on a set of sea-level and river discharge scenarios in the context of climate change (Aerts and Botzen, 2011)).
- In the United Kingdom, should global temperatures rise by 4°C, the annual average insured flood loss due to increased precipitation-based inland flooding could increase by 30%, resulting in an increase to the inland flood component of insurance premiums of approximately 21% (AIR Worldwide and UK Met Office, 2009).

*Risk diversification*

A large pool of diversified risk (independent and randomly-occurring losses) allows insurers to spread losses over a large number of properties and mitigate the potential for a large share of the pool to be affected by losses simultaneously. Other things equal, a smaller pool, or a pool with higher dependencies across the risks covered, will lead to higher premiums required by insurers (Schwarze and Wagner, 2007).

In the case of flood risk, building a sufficiently large pool of uncorrelated risks, including both high- and low-risk properties, is a significant challenge. Despite a high-level of uncertainty in assessing and modelling flood risk (see next section), there is a general understanding of which areas are – and are not – prone to flooding which will usually lead to adverse selection where there is no requirement for holding flood coverage (i.e. those interested in purchasing insurance will likely be limited to those facing higher levels of flood risk). Communities located in a riverine or coastal flood-plain are generally affected by floods more frequently than those that are located at a distance from (or elevation above) watercourses. Furthermore, the share of properties at high-risk of flooding is relatively small in most countries (see Table 3.2) – meaning that the vast majority of households and businesses in most countries face limited risk of flooding.

Table 3.2. **Estimates of the share of properties at high-risk of flooding**

| Country | Estimate |
|---|---|
| Australia | Riverine flooding: 7% of domestic houses[1]<br>1-in-100 year flooding: 160 000 homes[2] |
| Austria | Flooding (1-in-30): 150 000 exposed people[3]<br>Flooding (1-in-100): 350 000 exposed people[3]<br>Flooding (1-in-300): 650 000 exposed people[3] |
| Canada | Flooding (1-in-75): 13% of residential properties[3] |
| Czech Republic | Flooding (1-in-50): 9-10% of households[3] |
| Estonia | Flooding (1-in-50): 6 708 residents[3]<br>Flooding (1-in-100): 9 171 residents[3] |
| Germany | Flooding (1-in-50 to 1-in-200): 7.9% of households[4]<br>Flooding (1-in-50 or higher): 1.9% of households[4] |
| Ireland | Flooding: 300 communities identified as facing significant risk of damaging floods (based on index of hazard and consequences)[5] |
| Italy | Flooding and landslide (high-risk): 1.1 million residential buildings (9% of total)[9] |
| Latvia | Flooding (1-in-75): <1%[3] |
| Portugal | 2% of mainland Portugal displays high or very high vulnerability[6] |
| Russia | 7 400 settlements are located in "flood hazard areas"[3] |
| Spain | Flooding (1-in-100): 3.3% of population[3] |
| United Kingdom | Some degree of flood risk: 6 million properties (16.7%)[7]<br>Riverine and coastal flooding (1-in-75): 560 000 properties (England and Wales)[7] |
| United States | Riverine flooding (1-in-100): 4.9 million housing units[8]<br>Coastal flooding (1-in-100): 3.8 million housing units[8]<br>Coastal flooding (1-in-100): 16.4 million residents (5% of population)[3] |

Sources: [1] Allianz Australia Insurance Ltd. (2011); [2] Collins and Simpson (2007); [3] Country responses to an OECD questionnaire on the financial management of flood risk (2015); [4] GDV (2015a); [5] Office of Public Works (2012); [6] Costa et al. (2014); [7] Ramsbottom, Sayers and Panzeri (2012); [8] National Research Council (2015); [9] Swiss Re (2015b).

The difficulty in attracting low-risk households into a flood pool limits the diversity and size of the pool, and forces insurers to charge (even) higher premiums to high-risk households seeking coverage. Where flood insurance is offered as stand-alone coverage on an optional basis, there will generally be limited take-up of flood coverage from low-risk households. For example, a study in the United States found that some households outside the SFHAs (1-in-100 year risk of flooding) perceived themselves to be at no risk of flooding (GAO, 2014) – and, as noted, less than 1% of households in the 1-in-500 year flood zone are insured. A similar challenge was faced by an insurer attempting to introduce flood insurance in Winnipeg (Canada) as demand for coverage was found to be weak or non-existent in low-risk areas (Thistlethwaite and Feltmate, 2013). Some argue that this is the most important reason why flood insurance is not broadly available from the private sector in many countries (Swiss Re, 1998).

Another pressure on pricing for flood coverage is that insurers tend to respond to "accumulation risk" or "correlation risk" (i.e. the risk of facing losses across a significant share of an insured portfolio from the same event) by charging higher premiums (The Australian Government the Treasury, 2011). The tendency of flooding to impact large areas, such as a coastal or river floodplain, creates accumulation risk as a large number of losses occur simultaneously (Allianz Australia Insurance Ltd., 2011). An insurer providing coverage in a given region subject to correlated risks could therefore face a portfolio with a limited diversity of risks (i.e. a lack of mutuality). The Queensland floods in Australia in 2011, for example, led to losses across vast areas while historical floods in Australia affected even larger areas (Allianz Australia Insurance Ltd., 2011).

### *Uncertainty in quantification of potential exposures*

The low frequency of catastrophe events, combined with frequent changes in the level of assets at risk (due to continued economic development) and the uncertain impacts of a changing climate, makes catastrophe exposure particularly difficult for insurers to quantify. The frequent/continual losses in most lines of insurance business allow for statistical probabilities to be established with more certainty, whereas catastrophe events occur only infrequently, making accurate quantification significantly more difficult. This has led to reliance on complex catastrophe modelling for the setting of premiums able to cover expected losses. Catastrophe models use information on the probability of occurrence of events of varying magnitudes, the location, structural characteristics and vulnerabilities of assets-at-risk, and the level of insurance coverage of those assets to provide insurance companies with estimates of their exposure to different types of disaster events. The estimates are presented in terms of annual average loss and probability of loss exceedance (or exceedance probability) for a range of return periods that allow insurance companies to determine a price for providing coverage.

Flood risk poses a number of modelling challenges in terms of the scope of hazard modelling required, the impact on assets-at-risk and probability of occurrence:

- *Scope of hazard modelling*: Insurance policies will generally cover damage from flooding no matter what the cause (with some exceptions). However, the potential causes of flooding are numerous requiring significant investments in modelling to assess all possible scenarios. For example, a coastal city in a river delta could face flooding damage as a result of flash flooding, riverine flooding, groundwater flooding, coastal flooding or storm surge requiring insurers to model probabilities for many different types of both meteorological and hydrological risks to estimate their exposure to floods. Also, given that almost any area is susceptible to flash floods, modelling for flood risk is necessary for large areas, not just floodplains.

- *Impact on assets-at-risk*: Accurate flood risk assessments require detailed topographical information in order to project where water will flow and how fast, information on flood protection infrastructure and drainage systems and their relative effectiveness, detailed information on land-use in order to ascertain the level of water absorption (which is complicated where land-use patterns change rapidly, such as in fast-growing urban areas), as well as substantial information on the structure and its contents. The impact of floods on assets-at-risk depends on the level of water that actually reaches (and then penetrates) a given structure from a given precipitation or storm event. For example, a given area may be protected by a structural barrier which requires assessing the level of effective protection provided by that barrier and the potential for failure (which involves significant uncertainty). How much water reaches a given asset also depends on the amount of water that is captured by drainage systems or absorbed in the ground, which depends on land-use, type of vegetation as well as the level of antecedent wetness in the ground. The level of water penetration into a structure and subsequent damage will depend on the type of structure (wood vs. masonry, where wood is more susceptible to flooding (Ziehmann and Hilberts, 2015)), the elevation of the structure relative to the water level as well as the location of contents (particularly, electrical installations) within the structure. Very small differences in elevation (e.g. the height of a curb) can have important implications for the depth of water reaching a given asset-at-risk.

- *Evolving probability of occurrence*: Climate change is increasing the already significant level of uncertainty involved in understanding the frequency and severity of flood events (see Box 3.3). There is uncertainty with respect to both the ultimate level of greenhouse gas emissions as well as the impact of the build-up of greenhouse gases in the atmosphere on precipitation patterns, sea-level rise and storm generation. Climate feedback loops and tipping points exacerbate these challenges and are not clearly understood (Jotzo, 2010).

### Box 3.3 **Accuracy of hurricane forecasting**

North Atlantic hurricanes and tropical storms are among the most-modelled of all flood-related hazards given the significant levels of insured exposure, especially in the South-eastern United States, and the significant risk transfer that occurs for North Atlantic hurricane losses (including through international reinsurance and capital markets). However, hurricanes remain extremely difficult to predict given the number of factors that affect the generation of hurricanes including sea-surface temperatures and wind shears which are a function of periodic phenomena (such as El Niño) and the interactions between different atmospheric systems and the Atlantic ocean. On an annual basis, predictions are made on the number of hurricanes that will be generated, although given the complexity of weather systems, the level of accuracy of these forecasts is relatively low. Figure 3.4 shows the annual hurricane forecasts provided by the US National Oceanic and Atmospheric Administration (NOAA) relative to the actual number of hurricanes that were generated between 2002 and 2014.

Figure 3.4. **Annual hurricane predictions and actual hurricanes generated**

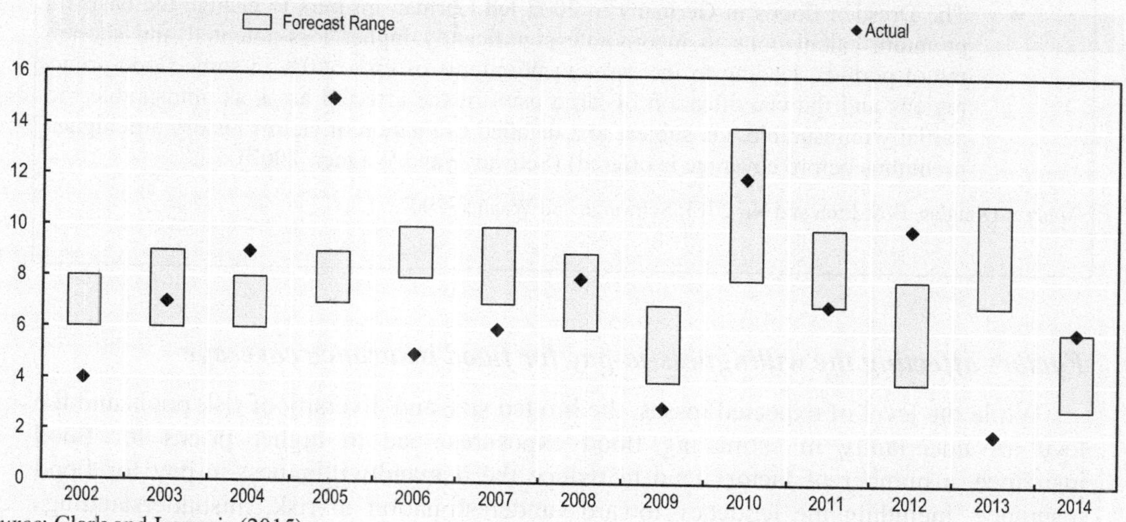


*Source*: Clark and Lummis (2015).

While catastrophe modelling (and continual improvements in the science (geophysics, meteorology, climatology, seismology, volcanology) behind such models) has greatly improved insurers' ability to quantify expected losses, the level of uncertainty related to expected disaster (and particularly flood) losses remains much more significant than in other lines of business (see Box 3.4). Insurance companies will tend to mitigate uncertainty in their estimates of expected losses by charging higher premiums. Research undertaken by Kunreuther et al (1995), based on a survey of underwriters, found that uncertainty in the understanding of a risk by the underwriter leads to significantly higher (1.43 to 1.77 times higher) premiums than the suggested pricing for a better understood risk. This is because an underestimation of risks can have significant implications for insurer solvency, leading insurers to account for this risk by adding an uncertainty premium. For example, Aerts and Botzen's (2011) analysis of the impact of various climate change scenarios on premiums for the coverage of flood risk in the Netherlands estimated that insurance companies would face a shortfall in reserves of almost 50% by 2030 if they wrongly set premiums in 2015 based on the expectation of a low sea-level rise scenario but were faced in reality with a high sea-level scenario (in practice, insurers would be able to increase premiums as the new scenario became evident, subject to any political or regulatory impediments to rate increases). The level of uncertainty is particularly high in developing and other countries where catastrophe models are not available (see section 4.2).

Box 3.4. **Post-event price adjustments**

The uncertainty inherent in estimating expected flood losses is evident in the significant fluctuation in insurance premiums for flood risk after flood events that seemed to have surprised the sector:

- After the Queensland floods and Cyclone Yasi in 2010-11 and subsequent floods and bushfires in 2011-12, premiums for home building insurance for some properties prone to flooding, cyclones or other natural perils in Australia increased by 400% (Douglas, Bowditch and Ni, 2013).
- The Dresden floods in Germany in 2002 led German insurers to change the basis of premium calculations to incorporate significantly higher loss-potential and shorter return periods, leading to increases in premiums of up to 60% in some flood-prone regions and the classification of large parts of the affected areas as uninsurable (or partially uninsurable, i.e. subject to a detailed evaluation of claims history and higher premiums before coverage is offered) (Schwarze and Wagner, 2007).

*Source*: Douglas, Bowditch and Ni, 2013; Schwarze and Wagner, 2007.

## *Factors affecting the willingness-to-pay for flood insurance coverage*

While the level of expected losses, the limited size and diversity of risk pools and the level of uncertainty in estimating flood exposures lead to higher prices for flood insurance, a number of factors tend to reduce the demand/willingness-to-pay for flood insurance, including the tendency towards underestimation of risk, misunderstandings about coverage and expectations of post-disaster compensation or financial assistance.

As a general rule, individuals (and businesses) tend to underestimate their exposure to disaster risks which reduces their willingness-to-pay for insurance coverage. The likelihood of being impacted by a low-probability event is systematically underestimated by individuals with some controlled experiments finding many individuals unwilling to pay anything for insurance coverage against low-probability events (McClelland, Schulze and Coursey, 1993). As noted above, a general understanding of the causes of flooding (e.g. proximity to river or coast) may exacerbate the underestimation of risk among low-risk populations not obviously exposed to these customary causes of flooding. Similarly, the construction of highly-visible protective infrastructure (such as the infamous levees in New Orleans which were overtopped during Hurricane Katrina) may give communities the impression that they are no longer at risk of flooding. For example, in a survey of residents in Grand Forks (United States) after severe flooding in 1997, the second most important factor for not purchasing flood insurance was a belief that dikes and other flood control measures would provide protection (Pynn and Ljung, 1999). Policyholders also tend to allow flood insurance coverage to lapse after a few flood-free years.

Homeowners and businesses insured against fire and storm damage may not be aware of any exceptions to coverage for floods or other natural disasters. For example, in Australia, there was a general backlash against the insurance industry following the 2011 Queensland floods from homeowners that were unaware that flood damage caused by riverine flooding was not covered in their home insurance policy. The expectation of government assistance after a flood (or other disaster) event is also likely a factor in reducing demand for flood insurance, even where such assistance has been historically limited (Browne and Hoyt, 2000; Michel-Kerjan, 2010; GAO, 2014). A number of countries noted that the expectation of government compensation was a significant challenge to insurance penetration (Russia, Latvia, Turkey, Portugal, United States). A study on post-disaster grants in the United States found a statistically significant (negative) relationship between the level of post-disaster assistance for a given area and the level of insurance coverage (see Box 3.5).

### Box 3.5. **The impact of financial assistance on insurance coverage in the United States**

Kousky, Michel-Kerjan and Raschky (2014) examined the levels of insurance coverage in a number of US coastal regions following the occurrence of a disaster event and the provision of financial assistance to affected communities. While they found that the provision of financial assistance had a limited (or even positive) impact on decisions to insure, higher-levels of average financial assistance to a given community (based on postal codes) had a negative impact on the amount of insurance coverage chosen by households in that community - and communities that received lower average levels of financial assistance generally saw an increase in the amount of insurance coverage chosen by households. The authors concluded that insurance requirements tied to the extension of financial assistance likely had the intended impact of increasing (or at least not decreasing) the number of households that chose to purchase insurance. Low levels of financial assistance may demonstrate the need for insurance coverage while higher levels of assistance may reduce the amount of insurance coverage secured as financial assistance is seen as an alternative to insurance.

*Source*: Kousky, Michel-Kerjan and Raschky, 2014.

### *The flood insurance market failure*

The combination of forces driving higher-prices and lower willingness-to-pay will often lead to a market failure in the private market for flood insurance. Low demand for flood insurance will reduce the size and diversity of the pool of risks (with limited participation from low risk households) leading to higher prices that further reduce demand (see Figure 3.5).

Figure 3.5. **Flood insurance market failure**

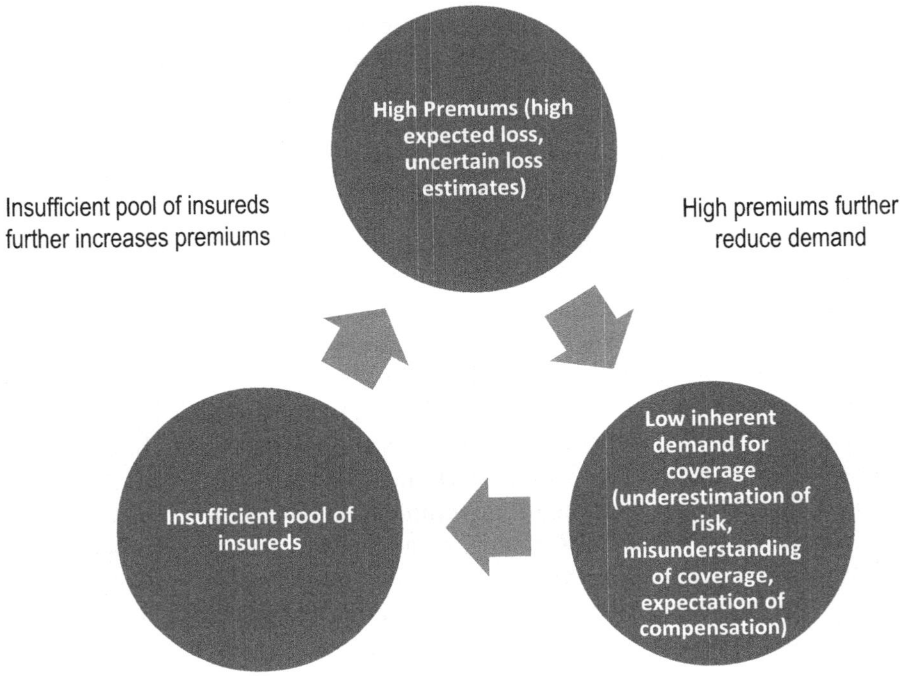


Low levels of insurance coverage in the event of a flood is likely to lead to greater pressure on governments to provide compensation (where such compensation is discretionary). Higher levels of government compensation is, in turn, likely to further reduce demand for insurance coverage (along with incentives for risk reduction, see next chapter). This has been termed the "disaster syndrome" (Kunreuther, 2000).

## Notes

1. This requirement applies to residential and commercial properties in Special Flood Hazard Areas (SFHA). The legislative requirement for mortgage lenders to ensure that the properties they lend against are protected by public flood insurance provided by the National Flood Insurance Program is long-standing although a proposed legislative amendment ("Flood Insurance Market Parity and Modernization Act") would also allow flood insurance from private insurers to be considered as meeting this requirement.

2. For European Union member states, any aid granted by a Member State or through State resources "which distorts or threatens to distort competition by favouring certain undertakings or the production of certain goods shall, in so far as it affects trade between Member States, be incompatible with the internal market" and must be notified to the European Commission. In general, aid "to make good the damage caused by natural disasters" is exempted from the notification requirements, as clarified through Commission Regulation No 651/2014 of 17 June 2014.

3. Earthquake losses also account for a larger portion of uninsured losses (34.7%) than would be expected based on the share of all losses caused by earthquake damage (28.4%).

## References

Aerts, J., and W. Botzen (2011), "Climate change impacts on pricing long-term flood insurance: A comprehensive study for the Netherlands", *Global Environmental Change*, 21. pp. 1045-1060.

AIR Worldwide and UK Met Office (2009), "The Financial Risks of Climate Change," *ABI Research Paper No. 19*, Association of British Insurers, London.

Allianz Australia Insurance Ltd. (2011), *Natural Disaster Insurance Review Submission*, 21 July.

The Australian Government the Treasury (2011), *Natural Disaster Insurance Review: Inquiry into flood insurance and related matters (September 2011)*, Commonwealth of Australia, *www.ndir.gov.au/content/report/downloads/NDIR_final.pdf.*

Bin, O. and C. Landry (2013), "Changes in implicit flood risk premiums: Empirical evidence from the housing market", *Journal of Environmental Economics and Management*, Vol. 65, pp. 361–376.

Browne, M. and R. Hoyt (2000), "The Demand for Flood Insurance: Empirical Evidence", *Journal of Risk and Uncertainty*, Vol. 20 (3), pp. 291-306.

Clark, K. and J. Lummis (2015), *Climate Change and Hurricane Loss: Perspectives for Investors*, Karen Clark and Company, Boston, June.

Collins, E. and L. Simpson (2007), "The impact of climate change on insuring flood risks", Presented to the Institute of Actuaries of Australia Biennial Convention, 23-26 September, Christchurch, New Zealand, *http://actuaries.asn.au/Library/1.a_Conv07_Paper_Collins%20Simpson_The%20impact%20of%20climate%20change%20on%20insuring%20flood%20risk.pdf.*

Costa et al. (2014), "National Flood Vulnerability Index", *Project CIRAC - flood risk and vulnerability mapping in climate change scenarios, http://siam.fc.ul.pt/cirac/outputs/relatorios/Vulnerabilidade/CIRAC_Vulnerability.pdf,* accessed 20 November 2015.

DEFRA (2013), *Securing the future availability and affordability of home insurance in areas of flood risk*, UK Department for Environment, Food and Rural Affairs, London.

Douglas, J., M. Bowditch and A. Ni (2013), *Affordability of Natural Disaster Insurance*, Australian Centre for Financial Studies, *www.australiancentre.com.au/sites/default/files/NewsDocs/D1P4%20-%20Affordablity%20of%20natural%20disaster%20insurance%20Justin%20Douglas%20Treasury.pdf*, accessed 6 October 2015.

Fédération française des sociétés d'assurances (2016), *Inondations: l'indemnisation des dommages en cas de catastrophe naturelle* (website), *www.ffsa.fr/sites/jcms/c_47622/en/inondations-lindemnisation-des-dommages-en-cas-de-catastrophe-naturelle?cc=p1_1534660*, accessed 23 March 2016.

Flood Re (2015), *How it works* (website), *www.floodre.co.uk/how-it-works-overview*, accessed 12 November 2015.

GAO (2014), *Flood Insurance: Strategies for Increasing Private Sector Involvement (GAO-14-127)*, Government Accountability Office (January), Washington, DC.

GDV (2015a), *Online-Serviceteil zum Naturgefahrenreport 2015: Tabellen, Grafiken und Karten*, German Insurance Association (Gesamtverband der Deutschen Versicherungswirtschaft), GDV, Berlin, *www.gdv.de/wp-content/uploads/2015/11/GDV-Naturgefahrenreport-2015-Unwetterbilanz-Versicherer-OnlineServiceteil-n.pdf.*

GDV (2015b), *Naturgefahrenreport 2015*, German Insurance Association (Gesamtverband der Deutschen Versicherungswirtschaft), GDV, Berlin, *www.gdv.de/wp-content/uploads/2015/11/Naturgefahrenreport_2015.pdf.*

Green, C. and E. Penning-Roswell (2004), "Flood Insurance and Government: "Parasitic" and "Symbiotic" Relations", *The Geneva Papers on Risk and Insurance*, Vol. 29 (3), The Geneva Association (The International Association for the Study of Insurance Economics).

Gupta, S.S. and R. Agrawal (2015), "Micro-insurance for Disaster Management – A Case of Afat Vimo in Orissa", *Journal for Studies in Management and Planning*, Vol. 1(5).

Guy Carpenter (2015), "Flood Insurance Risk Study: Reinsurance Study", *National Flood Insurance Program: Report to Congress on Reinsuring NFIP Insurance Risk and Options for Privatizing the NFIP (13 August)*, Department of Homeland Security, Washington, DC.

Insurance Bureau of Canada (2015), *The financial management of flood risk: An international review – lessons learned from flood management programs in G8 countries*, IBC, Toronto.

Insurance Europe (2012), *Insurance Europe key points for insurers regarding natural catastrophes in Europe*, Insurance Europe, 27 November.

Jametti, M. and T. von Ungern-Sternberg (2006), "Risk selection in natural disaster insurance: the case of France", *CESIfo Working Paper No. 1683*, March.

Jha, Abhas K. and Z. Stanton-Geddes (eds.) (2013), *Strong, Safe, and Resilient: A Strategic Policy Guide for Disaster Risk Management in East Asia and the Pacific*, World Bank, Washington, DC.

Jotzo, F. (2010), "Prerequisites and limits for economic modelling of climate change impacts and adaptation," *Environmental Economics Research Hub Research Reports*

*No. 55*, Environmental Economics Research Hub, Crawford School of Public Policy, The Australian National University.

Kousky, C., E. Michel-Kerjan and P. Raschky (2014), "Does Federal Assistance Crowd Out Private Demand for Insurance?", *Risk Management and Decision Processes Center Working Paper # 2014-04*, Wharton School, Philadelphia.

Kunreuther, H. (2000), "Strategies for dealing with large-scale and environmental risks", in *Frontiers in Environmental Economics*, Folmer, H. et al. (eds.), Edward Elger.

Kunreuther, H., et al. (1995), "Ambiguity and underwriter decision processes," *Journal of Economic Behavior and Organization*, Vol. 26, pp. 337-352, *http://opim.wharton.upenn.edu/risk/library/J1995_JEBO_HK_Ambiguity.pdf.*

Kunreuther, H., E. Michel-Kerjan and N. Ranger (2013), "Insuring future climate catastrophes", *Climactic Change*, Vol. 118, pp. 339-354.

Lamond, J. and E. Penning-Roswell (2014), "The robustness of flood insurance regimes given changing risk resulting from climate change", *Climate Risk Management*, Vol. 2, pp. 1–10.

Malagardis, A. (2015), "Building Institutional Capacity for the Development of Insurance Markets", Presented to the Global Seminar on Disaster Risk Financing: Towards the development of effective approaches to the financial management of disaster risks, 17-18 September, Kuala Lumpur, Malaysia, *www.oecd.org/daf/fin/insurance/global-seminar-disaster-risk-financing-2015.htm.*

Maccaferri, S., J. Carboni and F. Campolongo (2012), "Natural Catastrophes: Risk Relevance and Insurance Coverage in the EU", *EUR - Scientific and Technical Reports No. JRC67329*, Joint Research Centre, Ispra, Italy.

McClelland, G., W. Schulze and D. Coursey (1993). "Insurance for Low-Probability Hazards: A Bimodal Response to Unlikely Events", *Journal of Risk and Uncertainty*, Vol. 7, pp. 95-116.

Michel-Kerjan, E. (2010), "Catastrophe Economics: The National Flood Insurance Program", *Journal of Economic Perspectives*,Vol. 24 (4), pp. 165-186.

National Academies of Sciences, Engineering, and Medicine (2015), *A Community-Based Flood Insurance Option*, The National Academies Press,Washington, DC.

National Research Council (2015), *Affordability of National Flood Insurance Program Premiums: Report 1*, The National Academies Press, Washington, DC.

OECD (2015), *Disaster Risk Financing: A global survey of practices and challenges*, OECD Publishing, Paris, *http://dx.doi.org/10.1787/9789264234246-en.*

Office of Public Works (2012), *The National Preliminary Flood Risk Assessment (PFRA): Overview Report*, Office of Public Works (Ireland).

Pynn, R. and G. Ljung (1999), "Flood Insurance: A Survey of Grand Forks, North Dakota, Homeowners", *Applied Behavioral Science Review*, Vol. 7 (2), pp. 171-180.

Ramsbottom, D., P. Sayers and M. Panzeri (2012), "Climate Change Risk Assessment for the Floods and Coastal Erosion Sector", report for the UK 2012 Climate Change Risk Assessment, *http://randd.defra.gov.uk/Document.aspx?Document=CCRAfortheFloodsandCoastalErosionSector.pdf.*

Schwarze, R. et al. (2011), "Natural Hazard Insurance in Europe: Tailored Responses to Climate Change are Needed", *Environmental Policy and Governance*, Vol. 21, pp. 14-30.

Schwarze, R. and G. Wagner (2007), "The Political Economy of Natural Disaster Insurance: Lessons from the Failure of a Proposed Compulsory Insurance Scheme in Germany", *European Environment*, Vol. 17, pp. 403–415.

Seifert et al. (2013), "Influence of flood risk characteristics on flood insurance demand: a comparison between Germany and the Netherlands", *Natural Hazards and Earth Systems Science*, Vol 13, pp. 1691–1705.

Surminski et al. (2014), "Reflection on the current debate on how to link flood insurance and disaster risk reduction in the European Union", *Centre for Climate Change Economics and Policy Working Paper No. 184/ Grantham Research Institute on Climate Change and the Environment Working Paper No. 162*, Centre for Climate Change Economics and Policy, Leeds.

Swiderek, D. and J. Wipf (2015), *Aiding the disaster recovery process: The effectiveness of microinsurance service providers' response to Typhoon Haiyan*, Microinsurance Network, Luxembourg.

Swiss Re (2016), *The property protection gap in Latin America*, Swiss Reinsurance Company Ltd., Zurich.

Swiss Re (2015a), *Closing the gap: Flood protection for Bangladesh*, Swiss Reinsurance Company Ltd., Zurich.

Swiss Re (2015b), *The natural catastrophe protection gap in Italy: time for action*, Swiss Reinsurance Company Ltd., Zurich.

Swiss Re (2012), *Flood – an underestimated risk: Inspect, Inform, Insure*, Swiss Reinsurance Company Ltd., Zurich.

Swiss Re (1998), *Floods – an insurable risks*, Swiss Reinsurance Company Ltd., Zurich.

Thistlethwaite, J. and B. Feltmate (2013), *Assessing the Viability of Overland Flood Insurance: The Canadian Residential Property Market*, The Co-operators.

UNISDR (2014), *2014 Finland Peer Review – Building resilience to disasters: Implementation of the Hyogo Framework for Action (2005-2015)*, UNISDR, EC, OECD.

Wharton Risk Management and Decision Processes Center (2016), *Flood insurance around the world: Understanding residential flood insurance markets (beta version)*, *https://riskcenter.wharton.upenn.edu/flood-insurance-around-the-world/*, accessed 23 March 2016.

Wilby, R. and R. Keenan (2012), "Adapting to flood risk under climate change", *Progress in Physical Geography*, Vol. 36 (3), pp. 348–378.

Ziehmann, C. and A. Hilberts (2015), "Risk on the rise: managing inland flood risk", Webinar Presentation, 1 October, Risk Management Solutions Inc., *www.slideshare.net/RMS_News/risk-on-the-rise-managing-inland-flood-risk?mkt_tok=3RkMMJWWfF9wsRonuqjJZKXonjHpfsX%2F7uUpT%2Frn28M3109ad%2BrmPBy%2B3YIHWp8na%2BqWCgseOrQ8l14BV8S%2FRM0VrKA%3D*.

# *Chapter 4*

## Improving the insurability of flood risk

*This chapter provides an overview of possible approaches to improving the insurability of flood and their relative effectiveness based on the experience of many countries. This includes improvements in land-use planning, investments in risk reduction at the community and household levels as well as efforts to improve understanding of flood risk and the need for financial protection. The role of premium subsidies in encouraging demand for flood insurance is also examined, including some of the challenges in eliminating such subsidies over time.*

Chapter 3 outlined the general market failure in private insurance markets based on a number of factors that lead to an increase in the price of flood insurance premiums (size of expected losses, limited and correlated risk pools and uncertainty in the quantification of potential exposures) and a number of factors that reduce the willingness-to-pay for flood insurance (underestimation of risk, misunderstanding of coverage and expectation of government compensation). This chapter will provide an overview of the measures that have been put in place in countries to address these specific factors, including prevention and risk reduction measures aimed at reducing the frequency and/or severity of flood losses, improvements in the mapping of flood risk that should reduce uncertainty as well as a number of interventions aimed at supporting demand and willingness to pay. The effectiveness of various approaches to improving private insurance coverage of flood risk will also be discussed.

## 4.1 Investments in risk reduction

Investments in prevention to lower the probability of a flood event occurring or in mitigation to reduce the losses resulting from a flood event is a critical element in the financial management of flood risk. Such investments can also be particularly effective in the case of floods relative to other natural disaster risks (Botzen and van den Bergh, 2009) given the greater ability to protect communities and structures from water penetration through land-use planning, structural mitigation investments and household mitigation investments.

The types of prevention and mitigation investments can be grouped into three main categories: i) land-use planning and restrictions aimed at reducing the level of assets exposed to flood risk or reducing the impact of flooding through the use of natural mitigation measures (e.g. wetlands, mangroves); ii) community structural flood mitigation measures (e.g. flood defences, drainage systems) aimed at protecting particular areas against inundation; and iii) risk reduction at the level of individual properties (e.g. elevating a property). These types of investments are not mutually exclusive and should all be considered as part of a comprehensive approach to flood prevention and mitigation.

Analyses of the potential benefits of risk reduction in terms of reducing future losses have generally shown that risk reduction measures can create substantial benefits. The US Federal Emergency Management Agency has estimated, based on a review of 4 000 risk reduction programs, that the average cost-benefit ratio for investments in risk reduction is 1:4 (i.e. the benefits from risk reduction projects are 4 times the cost) (MMC, 2005). In the United Kingdom, the Environment Agency estimates that its capital investment in the flood and coastal erosion sector can achieve a whole life cost-benefit ratio of 1:9 or higher (Environment Agency, 2014).

Despite these potential benefits, there is some evidence of general under-investment in prevention and risk reduction. In the United Kingdom, an analysis by the Committee on Climate Change suggested that national government spending between 2011/12 and 2014/15 on flood and coastal erosion risk management was almost 20% below what the Environment Agency estimated as necessary to avoid increasing the number of households facing significant flood risk (Adaptation Sub-Committee Secretariat, 2014). However, this was recently addressed by a government commitment to increase investment at levels that are consistent with the Environment Agency's assessment of a long-term investment profile that maximises benefits in terms of reducing flood damage (Environment Agency, 2014b). A study by the International Federation of Red Cross and Red Crescent Societies estimated that USD 40 billion of investments in disaster preparedness, prevention and risk reduction could have reduced global economic losses by USD 280 billion in the 1990s (IFRC, 2001).

Many countries allocate significantly more funds to disaster response than risk reduction. For example, the US federal government spent an average of over USD 3 billion on disaster response annually between 1985 and 2004 compared to USD 195 million on disaster prevention during the same period (Healy and Malhotra, 2009). In the past two decades, approximately 87% of the estimated USD 107 billion provided as development assistance for disaster risk management was devoted to post-disaster response and reconstruction, and only 13% was devoted to risk reduction and other *ex ante* risk management measures (Keating, A. et al., 2014). In its review of natural disaster funding arrangements in Australia, the Productivity Commission (2014) recommended a significant shift in funding towards prevention (and away from recovery) (see Box 4.1).

Box 4.1. **Australia Productivity Commission findings on prevention vs. recovery**

In Australia, within the federal system, constitutional responsibility for natural disaster planning, mitigation and recovery sits with state and territory governments. The Australian Government has a role in assisting with the burden of relief and recovery after major disasters and in collaborating with all levels of government to strengthen communities' resilience to natural disasters and minimise their impact. The Australian government provides financial assistance directly to state and territory governments for prevention and preparation, response and – in particular – recovery (see Figure 4.1).

Figure 4.1. **Estimated Australian Government natural disaster expenditure**

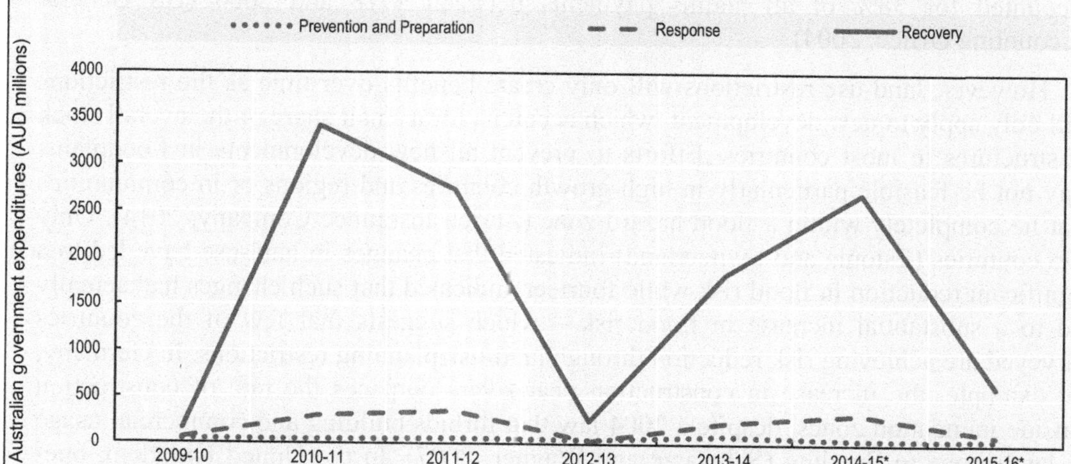


* Indicates estimates as the data was collected in mid-2014 and no adjustments since that time have been reflected in the data. Prevention and Preparation includes expenditures related to the National Emergency Management Projects, National Disaster Resilience Programme, National Bushfire Mitigation Programme, Betterment under the Natural Disaster Relief and Recovery Arrangements, National Flood Risk Information Project and Education training and research. Response includes expenditures related to the Crisis Coordination Centre, Counter-disaster operations of the Natural Disaster Relief and Recovery Arrangements, Defence Assistance to the Civil Community, National Aerial Firefighting Arrangements and Emergency Alert. Recovery includes expenditures related to the Natural Disaster Relief and Recovery Arrangements, Australian Government Disaster Recovery Payment, and Disaster Recovery Allowance

*Source*: Ward (2015).

**Box 4.1 Australia Productivity Commission findings on prevention vs. response (cont.)**

In 2014, the Australian Government asked the Productivity Commission to undertake a public inquiry into "the efficacy of current national natural disaster funding arrangements, taking into account the priority of effective natural disaster mitigation and the reduction in the impact of disasters on communities." The inquiry was specifically asked to examine "options to achieve an effective and sustainable balance of natural disaster recovery and mitigation to build the resilience of communities." Among the findings of the Productivity Commission (2014) is that post-disaster support to states and territories should be reduced while mitigation funding should be increased to AUD 200 million per year (which should be matched by the states and territories) – partly to ensure that states and territories are incentivised to invest in prevention. The Australian Government is consulting with states and territories on a modest and gradual approach to improving the balance between mitigation and recovery funding.

*Source*: Productivity Commission, 2014.

### *Land-use planning*

Land-use planning can have a significant impact on flood risk by reducing the level of assets at risk of flooding (i.e. restricting development in flood zones) and therefore slowing the accumulation of assets exposed to flood risk. Inappropriate land-use development can have a significant impact on losses. For example, in Thailand, the construction of industrial parks on former swamps that had been subject to regular flooding was a major factor in what became the largest ever insured loss from inland flooding (World Bank, 2012). In the United States, high-risk (repetitive loss) properties accounted for 38% of all claims payments between 1978 and 2004 (US General Accounting Office, 2004).

However, land-use restrictions will only create benefits over time as the restrictions will only apply to new development, which accounts for a small share of the overall stock of structures in most countries. Efforts to prevent all new developments in floodplains may not be feasible particularly in high-growth countries and regions or in communities that lie completely within a flood hazard zone (Zurich Insurance Company, 2014). Only two countries (Estonia and Switzerland) indicated that changes in land-use have led to a significant reduction in flood risk while fourteen indicated that such changes had actually led to a substantial increase in flood risk – which suggests that few of the countries surveyed are achieving risk reduction through land-use planning restrictions. In Germany, for example, the increase in construction near rivers outpaces the rate of construction outside inundation zones, despite a 2004 law that forbids building and commercial usage of land prone to flooding (Schwarze and Wagner, 2007). In the United Kingdom, one-third of the projected three million properties to be built by 2020 are expected to be located on coastal and river floodplains (Risk Management Solutions and Lloyd's (2008)). In Italy, the effectiveness of strong legislative requirements for assessing flood hazard in new construction has been limited by gaps in compliance and a number of amnesties provided for properties that were constructed without regard to flood hazard levels (Swiss Re, 2015).

Relocation of communities in high-risk areas is one (very expensive and disruptive) means of addressing past weaknesses in land-use controls. Relocation has been implemented in a number of countries in a post-disaster context. For example, New York

State implemented a buyout program in high-risk areas affected by Hurricane Sandy that compensated participating homeowners the full pre-storm market value of their home (Kaplan, 2013). To mitigate against the possibility of community deterioration (as a result of some residents relocating while others stay), the program offered a 10% compensation premium for blocks of residents that participated in the program on a collective basis (Siders, 2013). Relocation programmes have also been implemented in Australia and New Zealand. The town of Grantham in Queensland, Australia was relocated to higher ground following deadly flash flooding in 2011 through a voluntary land swap program (Wilby and Keenan, 2012). In New Zealand (while not related to flood), homeowners in a high-risk "red zone" encompassing areas prone to earthquake damage (due to thin crust, liquefaction, lateral spread and/or cliff collapse) were offered buyouts at near market value. Over 95% of eligible property owners participated in the program leading to the purchase of 7 300 properties in the red zone (Mitchell, 2015). In Switzerland, an industrial zone in Preonzo was relocated to a safer area in 2013 and a number of residential properties in a community close to Lucerne were removed due an unbearable risk of potential rock fall. In both cases, the costs were partially (but not completely) absorbed by the public sector.

In many cases, relocation programmes suffer from low levels of participation. This could be the result of households' attachment to the community (e.g. friends, family, employment, sentimental attachment) or potentially related to some of the other factors that reduce the willingness of households to protect against disaster risks (underestimation of risk, expectation of government assistance). A survey by Bukvic, Smith and Zhang (2015) of flood-affected households on the determinants of relocation decisions found that an increase in insurance rates (along with the potential recurrence of flooding) would provide a major incentive to relocate which suggests that risk-based insurance premiums could play a role in encouraging land-use controls in flood-prone areas (as long as households are forced to/or choose to remain insured). Cash settlement of insurance claims (instead of insurer-organised reconstruction) may be one means of facilitating relocation by providing claimants with significant payouts that would provide the means to relocate (The Australian Government the Treasury, 2011).

Public support for the purchase of high-risk properties could remove an important barrier to relocation although the purchase of high-risk properties could create legal risks by providing an implicit acknowledgement of the government's responsibility for protecting citizens against flood risk (see Box 4.2).

In many countries, land-use policy is a matter for local jurisdictions which often face pressures to allow development of (generally desirable) land near water sources. Local governments also tend to benefit most from development (through the generation of local tax revenues) while the costs that developments in risky areas create when a flood occurs may be shared more broadly. The benefits of development can also be realised in the short-term while the costs of risky development may only be incurred in the distant future. A study of disaster mitigation plans in Florida and North Carolina, for example, found that federal policies and incentives (such as the Community Rating System – see Box 4.3) had little influence on the prevalence of land-use actions in local mitigation plans as local authorities demonstrated a preference for meeting federal requirements through less problematic risk reduction actions such as emergency response and public awareness (Lyles, Berke and Smith, 2014).

### Box 4.2. **The role of liability in land-use planning**

Most governments would reject (and are careful to avoid any precedent for) a legal obligation to protect citizens against all disaster risks. For example, it would be reasonable to assume that the failure of an engineering structure, such as a dam, that results in flooding would create liability risks for the owner, operator and/or engineering contractor. A decision by a municipality to approve a development in a flood-prone area which is later flooded, on the other hand, would generally be considered as not subject to claims of liability directed towards the relevant decision-maker (Wilby and Keenan, 2012). However, in some countries, protection from liability for local planning decisions may not be assured. In Sweden, local councils have been found liable for flood damage in areas deemed unsuitable for development (Crichton, 2008). In the Netherlands, in the context of deliberations on the potential for a public-private partnership to provide natural disaster insurance, the Council of State (the supreme administrative court) advised against such a partnership as it could reduce the government's responsibility for the "habitability of land" (Jongjan and Barrieu, 2008), suggesting a potential obligation of the government that could have implications in terms of liability. The Productivity Commission (2014) review of natural disaster funding arrangements in Australia found some concerns among local governments about potential liability risk related to planning decisions and recommended that legislative protection should be provided to local governments for planning decisions taken "in good faith" (where such protections do not exist).

*Source*: Wilby and Keenan, 2012; Crichton, 2008; Jongjan and Barrieu, 2008; Productivity Commission, 2014.

In some countries, national governments can enact robust restrictions on land-use that must be applied by all local jurisdictions. For example, in Portugal, a national law forbids development in areas adjacent to rivers without pre-authorisation, even where the land is privately-owned. The size of the land subject to the restriction varies based on whether the river is affected by sea surge or tide or threatened by flooding (based on a 1-in-100 year return period) (OECD, 2014b). In Switzerland, a federal obligation for communities to undertake detailed hazard mapping has led to the incorporation of risk information into land-use planning in approximately two-thirds of Swiss communities (e.g. prohibitions on new building and reconstruction of destroyed buildings in high risk zones) (Federal Office of the Environment, 2015a).

National governments may also provide advice on and/or incentives for local land-use policies that reduce risk. In England, under the National Planning Policy Framework (Department for Communities and Local Government, 2012), the Environment Agency provides advice on flood risk to new developments and has been successful in influencing ultimate development decisions in some cases. In 2012/13, the Environment Agency's objections to local plans led to amendment in close to 99% of reported cases (DEFRA, 2013). National governments can also provide local governments with the necessary authorities for robust land-use planning. For example, in Denmark, national authorities passed legislation that specifically allows local governments to consider climate change and its risks in local planning decisions (OECD, 2013a).

Restrictions on access to insurance have been used in the United Kingdom as a means to influence local planning decisions where government reinsurance through Flood Re will not be available for developments constructed since 2009 (an approach established as part of the Statement of Principles agreement and to be maintained under Flood Re). This means that should insurers offer coverage to home and business owners in those developments, they will not have access to Flood Re reinsurance (and therefore may be unwilling to offer insurance coverage at all), which should incentivise better land-use planning (the Association of British Insurers (2009) has also published guidance for developers on how to increase the likelihood of securing private flood insurance coverage for new developments). Similarly, in Russia, insurance is not available for unauthorised structures in floodplains. In the United States, a significant objective of the National Flood Insurance Program has been to encourage appropriate local land-use development. NFIP insurance is only available to households in communities that have agreed to implement FEMA standards related to floodplain development (see Box 4.3).

### Box 4.3. **NFIP Community Rating System**

As a means of encouraging minimum standards of flood resilience in the United States, insurance coverage from the NFIP is only offered in communities (i.e. an administrative division with authority for enacting and enforcing development requirements) that agree to a set of flood management conditions, including building standards and floodplain management standards approved by FEMA, notably that structures be built above a base flood elevation level. These requirements have been estimated to have avoided USD 1.1 billion in flood losses per year and reduced the cost of flooding for an average residence in an SFHA from a 1-in-100 year event by over 65% (Sarmiento and Miller, 2006).

In addition, a Community Rating System (CRS) was established in 1990 as a voluntary program that communities may join. Communities that adopt recognised flood risk management practices (land-use planning and other risk reduction measures) above the minimum standards required for accessing NFIP insurance are awarded points. The points lead to a rating which in turn leads to a premium discount for households in that community (5% per rating level for those in an SFHA – up to 45%).

While only 5% of all communities have joined the program, those that have account for 68% of all policyholders (FEMA, 2015). Some communities have suggested that the burdens of entry into the program and of meeting the requirements for premium discounts have been a barrier to wider take-up, particularly among smaller communities with more limited risk management capacity (National Research Council, 2015). There is also some concern about the range of eligible measures, including both the ineligibility of some effective mitigation measures such as coastal sea walls and beach replenishment (Landry and Jahan-Parvar, 2011) as well as the eligibility of some measures (such as emergency response) which may be less effective in terms of reducing losses (Lyles, Berke and Smith, 2014). It has also been suggested that more points should be awarded for land-use planning given the relatively larger benefit of land-use restrictions in terms of reducing exposure (Lyles, Berke and Smith, 2014). However, an analysis of NFIP claims since 1998 found that households in communities with a high CRS rating (9 or above) submitted claims that were on average 13.5% lower than claims submitted by households in communities not participating in the program which suggests that the measures have been effective in reducing losses (Kousky and Michel-Kerjan, 2015).

*Source*: FEMA, 2015; National Research Council, 2015; Landry and Jahan-Parvar, 2011; Lyles, Berke and Smith, 2014; Kousky and Michel-Kerjan, 2015.

While restrictions on insurance in high-risk areas might enhance the effectiveness of land-use planning, the broad availability of affordable insurance coverage (or, alternatively, post-disaster government compensation) in communities facing significant flood risk could have the opposite effect by encouraging development in those areas and increasing the number of properties at risk.

Beyond establishing restrictions on development, land-use planning can also be used to increase the presence of natural flood protection mechanisms that can enhance water absorption and protect against storm surge. Natural flood defences, such as mangroves, wetlands, agricultural fields and green spaces, can provide significant benefits in terms of limiting the damage from coastal and inland flooding (OECD, 2016) and may be a more effective approach in areas that are already highly developed.

In the Netherlands, for example, the government initiated a "Room for the River Programme" that included the deepening of riverbeds and the establishment of 39 floodplains across major river systems (Orie and Stahel, 2013). The Netherlands has also established policies that encourage neighbourhoods to build floodable parks (Orie and Stahel, 2013). In Copenhagen, a "Cloudburst Management Plan", developed in response to extreme flash flooding in 2011, takes a comprehensive approach to leading rain waters to green and blue spaces that can absorb the water, including by designing streets to acts as urban waterways for the drainage of rain waters. Wetland and floodplain restoration is also being used to manage flood risk in other European river systems, such as the Danube (Ebert, Hulea and Strobel, 2009), and in Switzerland where many communities have widened riverbeds and predefined areas to be used as floodplains. In Australia, the Victorian State Government is buying land in order to restore it to natural floodplain functions.

The conservation and restoration of mangroves and coastal wetlands has also proven effective in providing protection against coastal floods and storm surges. Mangroves can reduce wave height by as much as 66% over 100 meters of forest (McIvor et al., 2012). For example, the "value" of coastal wetlands in the United States in terms of providing protection against hurricanes has been estimated at USD 23 billion per year (i.e. the estimated value of losses avoided as a result of the protection provided by wetlands) (Costanza et al., 2008). In Viet Nam, the protection of 12 000 hectares of mangroves (at a cost of USD 1.1 million) saved an estimated USD 7.3 million in annual dyke maintenance costs (Kay and Wilderspin, 2002).

### *Structural flood mitigation*

Structural flood mitigation measures, such as dams, levees and reservoirs as well as natural mitigation approaches such as the re-forestation of drainage basins, may be used to reduce flood risk by protecting areas from a given level of inundation and therefore reducing the frequency of flooding. Such measures may be the only cost effective approach to reducing flood risk in built-up areas located in floodplains (Productivity Commission, 2014). Improvements to urban drainage systems as well as the installation of permeable pavement (which is mandatory in the United Kingdom) can also be effective in improving water absorption capacity in urban areas. For example, improvements to the urban drainage systems in Mumbai could potentially reduce direct and indirect losses from a 100-year flood by 70% (Ranger et al., 2011). Due to the size of the investment needed, large structural mitigation investments tend to be most cost-effective where the value of assets protected is also very high (such as urban areas) (i.e. the benefits in terms of avoided losses justify the large cost of investment). In many major cities, including London, Tokyo and Bratislava, flood protection levels of 200-years or more have been established or are under development (OECD, 2014).

## Box 4.4. **The impact of structural flood mitigation investments: Some examples**

While individual flood events will usually differ from previous events, similar flood events affecting the same areas before and after an investment in flood mitigation as well as modelling techniques can be used to provide some insight into the benefit of structural mitigation in terms of reduced losses.

Similar flooding events affected parts of Germany, Austria, the Czech Republic and Switzerland in 2002 and 2013. Both events were caused by unusually high levels of spring precipitation followed by periods of intense rainfall that led to flooding as water could not be absorbed by saturated ground. The 2013 event involved higher reported water levels across most river gauges along the Danube and Elbe rivers and their major tributaries. However, the level of economic losses from the 2013 floods was relatively similar to the 2002 floods (EUR 12 to 16 billion relative to EUR 15 billion in 2002) despite the continued accumulation of assets and higher water levels. Cities such as Prague and Dresden managed to avoid major losses altogether in 2013 (relative to significant losses in 2002) through the installation of new structural mitigation measures. The number of levee failures declined to 5 major failures in 2013 relative to 13 in 2002. Similarly, an increase in the height of a dam in Munich by three meters likely saved the city from significant flooding in 2013 (Zurich Insurance Company, 2014).

The USD 125 billion in losses generated by Hurricane Katrina in the United States in 2005 remains the largest ever loss event globally from a tropical cyclone. Since that time, significant investments have been made in improving the structural protection around the city of New Orleans, which accounted for the largest share of damage from the hurricane, as a failure of the previous protection system was a significant driver of the level of damage. This has included the construction of a USD 14.5 billion storm surge defence system involving levees and floodwalls, pumping stations, canal closures and gated outlets as well as the adoption and enforcement of stricter building codes across the state of Louisiana. A simulation of the potential impact of a similar storm today, undertaken by a catastrophe modelling firm, estimated that losses across the 6 affected states (Alabama, Florida, Georgia, Louisiana, Mississippi, and Tennessee), assuming the storm surge defence system is able to protect the city of New Orleans, would decline substantially to approximately USD 71 billion, despite a 40% increase in the insured value of coastal property in Louisiana and a 36% increase in Mississippi (the two most affected states) (AIR Worldwide, 2015).

Modelling techniques have also been used to estimate the benefits of investments in coastal defences in the city of Hamburg since a storm surge in February 1962 reached 5.7 meters above sea level, killed 318 people and caused losses equivalent to EUR 1.6 billion (in current values). Flood protection infrastructure was reinforced and raised to protect against a storm surge of 8.0 meters above sea level at a cost of approximately EUR 2.2 billion. According to calculations by a reinsurance company, these investments have protected the city from coastal flooding (at levels above the level reached in 1962) four times since then resulting in savings of EUR 17.5 billion in losses not incurred (taking into account inflation and increasing asset values) (Munich Re, 2012).

*Source*: Zurich Insurance Company, 2014; AIR Worldwide, 2015; Munich Re, 2012.

Structural flood mitigation investments are a common measure for addressing flood risk in most countries and are generally considered effective (see Box 4.4). Almost all surveyed countries (seventeen of twenty) indicated that structural mitigation investments have led to a reduction in flood risk with some respondents suggesting that it is likely the largest contributor to reducing flood risk in their country. In Ireland, for example, an estimated 6 000 properties have been protected since 1995 with structural mitigation investment expected to lead to a reduction of EUR 1 billion in exposure over the life of the investments. In Switzerland, EUR 25 million in investments in structural measures to protect the commune of Buochs on Lake Lucerne against flooding from the Engelberger Aa river prevented an estimated EUR 150 million in damages from flooding in 2005 (Federal Office of the Environment, 2015b). In Portugal, new drainage infrastructure in Lisbon is being assessed as a possible solution to high-levels of flood risk. In Australia, mitigation investments are seen as a possible means of balancing the demand for development in highly-desirable waterfront locations and in the outskirts of urban areas with the need to manage flood exposures.

However, many countries noted significant barriers to implementing structural mitigation investments as a strategy for the financial management of flood risk. Large structural investments are costly and a number of countries noted significant challenges in terms of accessing the needed public funding (both for initial construction and ongoing maintenance). The time horizon of political cycles makes the avoidance of potential losses at some future time a less attractive rationale for the use of scarce public investment resources. In the case of the Czech Republic, securing resources for mitigation investments outside of large urban areas was identified as particularly challenging. In built-up areas, there are additional challenges related to securing community support for investments that – while providing enhanced protection against floods – limit direct access to shorelines or create blots on the landscape. The long-term effectiveness of structural mitigation was also identified as a potential challenge in light of the uncertain impacts of a changing climate on flood risk (see Box 4.5).

### Box 4.5. **The design of structural mitigation investments in a changing climate**

In many cases, the life of large mitigation investments will span well beyond the 21st century and therefore such structures are almost certain to be tested against the impacts of climate change. Extreme but plausible scenarios, such as a global average temperature increase of 6°C (the approximate average projected temperature increase for the year 2100 in a scenario of increasing carbon dioxide emissions), could have significant implications on planning assumptions over long time horizons. For example, in the Netherlands, a 6°C scenario could lead to mean rise of (non-storm surge) sea levels of 4 meters along the coast in 2200 (Vellinga et al., 2009). Given the significant costs that usually come with attempts to increase the protection level of existing mitigation infrastructure – and the significant costs that could be incurred as a result of a breach of that infrastructure – such extreme scenarios may need to be considered in the design of current investments.

A number of countries are implementing such an approach. For example, the Netherlands Delta Committee, tasked with providing advice on long-term flood protection in the context of climate change, considered high-end climate scenarios out to 2200 in its assessment. The Thames Estuary 2100 study also used a "high-plus-plus" climate scenario in its assessment of options for providing flood protection for London out to 2100 (Wilby and Keenan, 2012). In Germany, the *landers* of Bavaria and Baden-Württemberg have introduced a "climate surcharge" into the design of all flood-related structures (i.e. the (calculated) design flood discharge value (projected flood) is augmented by 15 % to account for climate change).

*Source*: Vellinga et al., 2009; Wilby and Keenan, 2012.

 

Governance was also identified as a challenge to effective structural mitigation of flood risk, including the difficulty in coordinating across the various agencies involved. The difficulties in ensuring coordination among agencies are exacerbated where river basins and/or exposed communities cross administrative lines. An assessment by the OECD (2014a) of resilience to floods in the Seine basin in Île de France found that institutional fragmentation between administrative levels has been an impediment to building resilience against flood risk.

In a few countries, the availability and affordability of insurance has been a consideration in decisions on investments in structural mitigation measures (although only 30% of respondents to the survey identified any significant relationship between the two issues). For example, in Australia, a range of specific mitigation investments have been made by the government to reduce the potential for loss and support effective private insurance coverage in areas severely affected by the 2010-11 Queensland floods. The relationship between insurance availability and mitigation was also at the core of an agreement between the UK government and insurance sector that preceded the establishment of Flood Re (see Box 4.6).

### Box 4.6. **Investments in mitigation and insurance availability in the United Kingdom**

Prior to the establishment of Flood Re in the United Kingdom, the insurance industry and the UK government had a long-standing arrangement for the provision of flood insurance coverage by the private insurance industry (known initially as the "Gentleman's Agreement", later replaced by the "Statement of Principles"). Initially, the arrangement guaranteed the availability of flood insurance as part of bundled household insurance coverage for all residential properties. In 2002, however, the arrangement was limited to providing a guarantee of the availability of coverage for all residential properties facing a level of flood risk below 1-in-75 years. In 2005, the Statement of Principles extended insurers' commitment to provide coverage for residential properties (and small business) where the flood risk level was expected to improve to below 1-in-75 years within five years as a result of investments in flood protection. In areas with higher levels of flood risk, the availability of insurance coverage was not guaranteed and would be considered on a case-by-case basis. The Statement of Principles was extended in 2008 for a last 5-year period and was amended to exclude any property built after 1 January 2009, on the basis that no further development should occur in areas with flood risk levels above a 1-in-75 year return period (DEFRA, 2013).

These agreements aimed to ensure broad availability of flood coverage from the private insurance industry, supported by a commitment from the government to implement structural and non-structural measures (flood maps, flood defences and land-use planning) to manage the level of flood exposure – and have often been identified as a best-practice in terms of public-private collaboration on the financial management of flood risk (e.g., Thistlethwaite and Feltmate, 2013; Swiss Re, 2012). They provided the government with a means for ensuring broad insurance coverage of private flood losses and the insurance sector with a means to compel government into making sufficient investments in flood mitigation. However, over time, dissatisfaction with the level of government investment (along with concerns about ensuring a level-playing field relative to new insurance market entrants not subject to the agreements) played a role in the replacement of the agreement through the establishment of Flood Re. Flood Re will operate with a continued government commitment to investments in flood defences and appropriate land-use controls (through a letter of comfort that will be provided by government) (Surminski and Eldridge, 2014).

*Source*: DEFRA, 2013; Thistlethwaite and Feltmate, 2013; Swiss Re, 2012; Surminski and Eldridge, 2014.

Investments in structural flood mitigation should lead to reductions in insurance premiums for private property within communities benefitting from enhanced levels of protection. For example, in Australia, one insurance company announced that premium reductions of 30 - 80% could be expected once the construction of a new levee was completed (Suncorp Insurance, 2013). These premium reductions (even expected future reductions) could be considered a potential source of funding for mitigation. Local governments investing in mitigation could consider recouping some of the expected savings that will benefit their residents through specific charges/taxes. For example, a community that invests in mitigation measures that lead to a NFIP premium discount for policyholders within the community under the Community Rating System could conceivably impose a tax for some or all of the amount of the expected premium discount in order to finance those investments.

Insurance companies themselves might also be a potential source of financing for structural mitigation. Insurance companies manage more than USD 28 billion in assets on behalf of policyholders and third parties (OECD, 2015b) and have a self-interest in supporting investments in resilience as a means of reducing losses over time. Governments could capitalise on this potential demand from insurance companies for supporting investments in resilience by issuing bonds for the express purpose of financing such investments. This emerging bond type, known as "resilience bonds" or "municipal adaptation bonds", is similar to the concept of "green bonds" whereby the issuer commits to use the funds for specific "green" purposes. In fact, such investments are usually considered to be an eligible use of proceeds from green bonds. Investment in climate change adaptation is included as an eligible use of funds under the Green Bond Principles. The Climate Bonds Standard and Certification Scheme, which aims to establish standards for "green" municipal bonds, includes investments in flood mitigation as an eligible investment that can be financed with the proceeds of a green municipal bond (US Green City Bonds Coalition, 2015). The City of New York announced in 2014 that it intended to issue ("green") bonds for the specific purpose of financing projects that would boost resilience to climate change (Owens, 2014). Green bonds have also been issued by Nederlandse Waterschapsbank (NWB Bank) to finance loans to Dutch water authorities for water management measures, including flood protection measures (Kidney, 2016).

Insurance companies are also significant investors in infrastructure. However, the potential for private investment in resiliency projects is limited by the challenge in structuring an approach that provides the investor with returns over time. Unlike toll roads or airports, structural mitigation projects such as flood barriers do not generate future revenue with which to repay investors and are therefore usually financed by public funding. One approach put forward to address this barrier is to link investments in resilience to pre-defined rebates on catastrophe bonds that could be used to fund the project costs (Vajjhala and Rhodes, 2015) - although a significant increase in interest in catastrophe bond issuance by public agencies would be a prerequisite. Regulators may want to consider whether the capital or liquidity treatment of investments in resilience bonds or resiliency projects is appropriate given the potential benefits of investments in structural mitigation for reducing exposure to flood losses.

Careful consideration also needs to be given to some of the potential negative impacts of flood protection infrastructure. Structural defences can (and do) fail with significant consequences. While the number of levee failures during the 2013 Central Europe floods were lower than in 2002, 19 levees still failed in Saxony alone, including 5 major breaches (Zurich Insurance Company, 2014). Losses would reach an estimated

USD 105 billion in the case that the New Orleans' storm surge system failed under the hypothetical Hurricane Katrina-like scenario described in Box 4.4 (AIR Worldwide, 2015). The flood mitigation infrastructure itself can also increase losses from flooding if it is damaged in the event and/or it slows the dispersion of floodwaters after the event (Crichton, 2008).

### *Household risk reduction*

In addition to investments in risk reduction to protect flood-prone areas, there are a number of measures that can be taken at the level of individual households to either limit the damage when flooding occurs or prevent inundation altogether. These include elevation of structures above flood water levels, elevated curb stones to prevent water entry from smaller events, reinforcement of foundations and cladding to avoid structural damage from fast-moving waters and/or debris, moving building contents (and particularly electrical installations) above flood water levels (either temporarily in the event of a flood or permanently), dry flood proofing to make areas below flood water levels water tight and temporary or permanent flood walls (ranging from sand bags to free-standing concrete barriers). Some of these measures are extremely costly for existing structures. For example, raising the ground floor of an existing building above floodwater levels could cost GBP 30 000 (or 35-50% of a typical home value in the United Kingdom) (Risk Management Solutions and Lloyd's, 2008) and may not be possible if the existing ceiling is too low. Temporary measures, such as the placement of temporary flood barriers may be more cost-effective than some of these permanent measures (depending on the frequency of flooding) (Wilby and Keenan, 2012).

A number of empirical assessments of the effectiveness of household risk reduction investments have found significant benefits in terms of reducing losses from floods. A study on the impacts of the Elbe River floods in Germany in 2002 found that flood adapted buildings faced 46% less damage to structures and 48% less damage to contents; flood-adapted interior fittings reduced damage to buildings and contents by 53%; and placing utility and electricity installations on higher floors reduced flood damage by 36% (Kreibich et al., 2005). In the United States, elevating a structure has been found to reduce claims as a share of building costs by 16% to 18% (at a cost of approximately USD 50 000 to USD 70 000 to elevate an existing home) (Kousky and Michel-Kerjan, 2015). Buildings constructed to meet stricter NFIP requirements after 1975 have been found to face six times less damage than those built pre-1975 (Pasterick, 1998).

Damage prevention efforts by residents in Cologne (Germany) were found to have been the primary factor in reducing damage from two floods with similar floodwater levels in 1993 and 1995 from EUR 65 million to EUR 30 million (Fink, Ulbrich and Engel, 1996). The International Commission for the Protection of the Rhine (2002) estimated that investments in long-term mitigation (e.g. protective water barriers and, particularly, the replacement of oil heating installations located in basements with gas heating installations located under ceilings) by households and firms could reduce monetary damages by 80%. Modelling of hurricane damage in the United States, comparing current building standards relative to legacy standards, found that losses could be reduced by 34% to 61% (depending on the state) if all structures were brought to current building standards (Kunreuther et al, 2008). High-resolution flood models have found significant changes in the level of damage to a given structure (10x or more) based on small variations in a building elevation or level of protection (R. Muir-Wood, 2014, personal communication, 21 August). A broad assessment of the cost-benefit ratios for household flood mitigation measures in 34 flood-prone developing countries found an average of 1:60 for constructing a one-meter wall around houses and 1:14.5 for elevating houses in flood-prone regions (Kunreuther and Michel-Kerjan, 2013b).

Despite these benefits, most studies that have examined the willingness of households to invest in flood mitigation have found little interest in such investments. A survey of US residents along the Atlantic and Gulf coasts in 2006 (after the record 2005 hurricane year) found that 83% of respondents had not taken any steps to fortify their homes (Goodnough, 2006). Surveys of earthquake-prone homeowners in California in 1989 found similar levels of inaction (only 5-9% of respondents had adopted any loss reduction measures) and little change over 15 years of attempts to increase public awareness of earthquake risks (Palm et al., 1990). Even where households have made an initial investment in purchasing protective materials, they may not take the additional step of installing those materials. For example, a survey of residents in coastal New York counties during Hurricane Sandy found that less than half of those that had purchased storm shutters had actually installed them before the hurricane (Meyer et al., 2013).

The reasons behind households' reluctance to invest in risk reduction are likely similar to the reasons behind households' limited willingness-to-pay for insurance coverage outlined in Chapter 3 (low risk awareness, expectation of government assistance, etc.). However, households' reluctance to invest in protecting themselves is likely to be exacerbated in the context of risk reduction investments that are generally more costly than an annual insurance premium – particularly in countries with high-levels of mobility where homeowners sell their homes more frequently. Household decision-makers may also exhibit myopic behaviour where they focus solely on the potential benefits of an investment over a short period of time, such as 2-3 years, and therefore underweight the benefits of mitigation investments over the longer-term (Kunreuther and Weber, 2014).

A number of countries have implemented funding programmes (either temporary or permanent) to support flood protection at the household level. National and local authorities in England provide funding for property-level mitigation measures (up to GBP 4 750 is available per household from the national government) (DEFRA, 2013). The government of Alberta (Canada) established a program to provide funding support to residents to invest in self-protection after the 2013 floods in Calgary (Stelmakowich, 2013). In the United States, federal public funding is available for property-level risk reduction both pre- and post-disaster. The pre-disaster programmes support building elevation, relocation, demolition and rebuilding for all types of properties and flood-proofing for commercial structures. A portion of the pre-disaster programme funding comes from the National Flood Insurance Fund and therefore from insurance premiums. Funding has varied significantly from year-to-year and demand for the programs has been well-above capacity (National Research Council, 2015). The federal post-disaster programmes are triggered by a presidential disaster declaration and respond to applications from states who disburse the funds to local governments for use at individual properties. Homeowners with a NFIP policy can also access up to USD 30 000 in "Increased Cost of Compliance" payments as part of a post-flood claim to bring a structure into compliance with current building standards, including base flood elevations (where applicable) (Kousky and Kunreuther, 2014). As in the case of structural mitigation investments for communities, there may be a case for targeting funding programs towards areas where there are difficulties in terms of insurance availability or affordability (National Research Council, 2015).

There may also be opportunities to use post-disaster funding for mitigation to leverage funding from insurance companies for risk reduction. Insurance companies will rarely pay for risk reduction improvements (or "betterment") of flood damaged properties covered by insurance (Wilby and Keenan, 2012) as resilient reinstatement of a damaged property can cost 40% or more than standard reinstatement (Wassell et al., 2009). Public funds, that might normally be available for post-disaster mitigation investments, could potentially share the costs of betterment in situations where insurance companies would not normally assume such costs. It has been suggested that post-disaster grants to households for resilience measures after floods in 2013-14 in the United Kingdom could have been made more effective in terms of reducing risks if the use of the grants was coordinated with insurance companies involved in rebuilding insured properties (which might also facilitate future premium reductions) (CII New Generation Programme Claims Group, 2015).

The attractiveness to households of risk reduction investments would become more obvious if such investments lead to (future) discounts in the premium paid for flood insurance. For example, a survey of homeowners in the Netherlands found that two-thirds of respondents would be willing to invest in water barriers, about a fifth would be willing to install water-resistant floor types and a quarter would be willing to move central heating installations to upper floors if such investments would lead to a reduction in their insurance premium (Botzen et al., 2009). In order to provide such discounts, insurance companies would need to be sure that the risk reduction investments would be effective in reducing future losses, which would require some assurance that the measure will be properly implemented and maintained. It may also be much more difficult for an insurance company to quantify the reduction in exposure expected from non-engineered small-scale measures (Ball, Werritty and Geddes, 2013). These types of concerns may be a factor in the relatively low number of countries where premium discounts are commonly available for risk reduction measures. Only eight of the twenty-seven surveyed countries indicated that premium discounts related to mitigation investments were possible and generally only in some cases or for some lines of business (such as commercial). A survey of German insurance companies found that only 14% rewarded voluntary private risk reduction measures (Thieken et al., 2006).

Some countries have implemented initiatives to help translate mitigation investments into meaningful assessments of reductions in losses. In Germany, a programme has been established to allow households in flood exposed areas to obtain a flood resilience certificate that is based on an extensive risk assessment by authorised experts (OECD, 2015a). Property owners can seek a flood pass ("Hochwasser Pass") that provides an assessment of the individual flood risk for a given structure and particular mitigation measures that could be implemented to reduce that risk. The flood pass can also be used to support access to insurance for properties that can demonstrate acceptable levels of flood risk. A similar approach, based on the Energy Performance Certificate required to attest to a structure's energy use and cost in the event of sale or lease of a property, has been proposed in the United Kingdom (CII New Generation Programme Claims Group, 2015). In the United States, specific guidance on mitigation has been developed by FEMA and some measures (including elevation, wet-proofing and permanently moving contents to a higher level) can lead to lower NFIP premiums (National Research Council, 2015). Under the *Biggert-Waters Flood Insurance Reform Act of 2012*, FEMA is required to examine the potential for other measures to lead to premium discounts. In the United Kingdom, a standardised "flood risk report" is made available by the government for use by surveyors in reporting on flood resistance and resilience measures implemented for a given property (DEFRA, 2013). In Florida, legislation has been introduced requiring insurance companies to provide discounts, rate differentials or reduced deductibles for properties where mitigation investments demonstrated to reduce wind losses have been implemented (Warner et al., 2009). In Australia, the government has provided funding for an inspection scheme to undertake building assessments on multi-residence structures in North Queensland where insurance affordability has been a challenge (The Australian Government the Treasury, 2015a).

Premium discounts are also important for reducing the moral hazard that insurance coverage might otherwise create. While households will always have an interest in protecting their belongings from flood damage, the financial incentive for risk reduction measures will not be as strong if the implementation of such measures has no impact on the cost of insurance. That said, there is some evidence from Germany and the United States that insured households have a higher tendency to implement self-protection measures than uninsured households when faced with potential flooding, despite the lower losses insured households would be expected to bear (Thieken et al., 2006 and Hudson et al., 2014).

Another potential approach to improving the incentives for household-level mitigation would be to provide financing for the costs of mitigation based on the expected future reductions in premium levels. For example, a homeowner with limited current financial capacity to fund mitigation investments should be able to seek a loan for those investments as long as the investments will lead to future premium discounts that could finance the repayment of the loan. As an example, one study estimated that a loan of USD 25 000 to elevate a home (with a term of 20 years and an interest rate of 3%) would cost USD 1 680 per year to service but lead to a reduction in premiums of close to USD 3 500, creating net savings of USD 1 800 per year (Kunreuther and Michel-Kerjan, 2013a). Such loans could be extended by the private or public sector, could be included as a package with flood insurance, and/or could be multi-year and transferable to purchasers of a given home (in order to provide an easier repayment schedule for large investments) (Michel-Kerjan, 2010; Kunreuther and Michel-Kerjan, 2013a; National Research Council, 2015). It could also be part of a premium subsidy provided to homeowners in high-risk areas to support the affordability of flood insurance (see section below).

Tax incentives could be another means of encouraging investments in mitigation although such approaches tend to disproportionately benefit those in higher income tax brackets. In the United States, a legislative proposal has been introduced that would provide a tax credit of up to USD 7 500 for qualifying mitigation expenses by individuals and SMEs that hold an NFIP policy and benefit from subsidised premiums (National Research Council, 2015). Another proposal being considered in the United States is to create a Disaster Savings Account that would allow homeowners to contribute up to USD 5 000 annually (pre-tax) to save for uninsured damages and/or investments in a list of qualifying risk reduction measures. Amounts withdrawn for qualifying expenses would not be taxed (National Research Council, 2015).

### *The benefits of a comprehensive approach to risk reduction*

The use of land-use planning, large-scale structural mitigation and household risk reduction measures to address flood risk need be considered in a holistic way to be effective. For example, structural investments risk being ineffective if not supported by appropriate land-use controls. Without appropriate land-use controls, investments in flood defences could encourage development of newly-protected areas which could increase losses were the defences to be breached. One of the respondents to the survey (Turkey) noted that efforts to reduce flood risk through significant investments in structural flood mitigation were being undermined by land-use changes that continually expanded the land area at risk through new developments (and also increased the demand for further structural protection).

A combination of measures can have a significant impact on future exposures, even taking into account the increasing hazard levels expected to accompany climate change. For example, a case study for the North European coast estimated that a combination of sea flood defences and individual property measures could reduce the 75% increase in losses expected with a 30 cm rise in sea-levels by 95% (Risk Management Solutions and Lloyd's, 2008). In Florida, an estimated 40% of expected losses accompanying a high climate change scenario could be avoided through the construction of levees, better management of vegetation and changes to building standards for houses and apartments (Swiss Re, 2010).

The objective for flood risk management planning should be to find an optimal portfolio of the flood risk management tools available, weighing the costs and benefits of the different investments and taking into account the cumulative benefits from different types of tools when combined. One approach might be to consider different possible flood risk reduction investments as a portfolio of investments and aim to optimise the return (and minimise risk) across a set of possible flood risk reduction investments (Aerts et al., 2008).

To support an integrated approach to flood protection, post-disaster assistance for "betterment" might benefit from allowing flexibility in the use of funds. In the United Kingdom, some local councils allowed households to pool funds received through Repair and Renew Grants provided after 2013-14 flooding to finance larger flood defence measures seen as more effective in providing protection at the community-level (CII New Generation Programme Claims Group, 2015).

## 4.2 Mapping and modelling of flood risk

As noted in Chapter 3, a key challenge to the insurability of flood risk is impediments to the accurate assessment of flood exposures, including the complexity of modelling flood risk as well as the uncertain impacts of climate change on precipitation and storm patterns going forward. Accurate risk maps that provide estimates of frequency and assessments of impacts on structures for all types of floods are critical for land-use planning and risk reduction investments, and underpin the pricing of insurance premiums. In a number of countries, a lack of high-quality maps that provide an up-to-date assessment of the level of flood risk by geographical area has been a significant impediment to effective financial management of flood risk as well as private insurance coverage of flood risk (see Box 4.7).

Flood hazard maps are usually prepared by public authorities and/or commercial vendors and provide information on flood probability based on the extent of potential flooding, water levels and/or flow velocity under different flood scenarios. These types of maps are tailored to the needs of public land use planning, where one application may be the establishment of restrictions on new development in high hazard areas. However, these types of maps are insufficient for the purpose of insurance coverage, where information on potential consequences of flooding is also necessary. Flood risk maps that provide information on the potential consequences of flood hazards based on information on structures located in inundation zones are more useful but less commonly available. At the time of writing, scenario-based catastrophe models have been developed by the (re)insurance sector and are also available from vendors for a few Latin American and Asian countries (Argentina, Mexico, Ecuador, Paraguay, Papua New Guinea, South Korea, Viet Nam and New Zealand). Fully probabilistic flood catastrophe models, which provide information on potential damages (including insured damages) at different levels of probability, are only available from vendors for a few countries, such as the United States (including storm surge for the Gulf and East coasts), Canada, Austria, France, Poland, Switzerland, Belgium, Netherlands, Hungary, Czech Republic, Slovakia, Germany, the United Kingdom, Mexico, Brazil, Indonesia, and Australia.

### Box 4.7. **Mapping challenges: Canada, Australia and the United States**

A common challenge in countries facing diverse levels of flood risk across their territories is the maintenance of consistent, high-quality and updated flood risk maps:

- In Canada, the quality of maps developed by conservation authorities and municipal, provincial and federal governments has been a significant impediment to private insurance coverage. The flood maps: i) do not provide sufficient information on the location and cost of potential damage or the frequency of flooding; ii) use different flood return periods for different regions within Canada (e.g. 1-in-100 in Alberta, 1-in-250 in British Columbia and various return periods in Ontario); iii) are outdated and therefore do not capture changes in urbanisation or climate change impacts; and iv) do not generally include flash flooding (Thistlethwaite and Feltmate, 2013 and Insurance Bureau of Canada, 2015 ).
- In Australia, a submission from Allianz Australia Insurance Ltd. (2011) to the Natural Disaster Insurance Review indicated that, with the exception of New South Wales, flood maps made available by local councils were not of sufficient quality to allow them to offer insurance coverage.
- In the United States, consultations with risk modelers found limited confidence in the reliability of flood maps prepared by FEMA. Among the findings from these consultations were that the maps: i) are based on less-sophisticated methodologies than are currently available; ii) underestimate base flood elevations (and therefore flood risk to structures); and iii) are unable to account for important factors such as changing climate patterns, building stock growth and the interaction between different bodies of water (GAO, 2014). A US GAO (2008) report suggested that 50% of FEMA flood maps were more than 15 years old and 8% were 10-15 years old in 2008. More recent statements by FEMA suggest progress, although 40% of maps are still considered in need of review and 10% are considered out-of-date (Simpson, 2014).

*Source*: Thistlethwaite and Feltmate, 2013; Insurance Bureau of Canada, 2015; Allianz Australia Insurance Ltd., 2011; GAO, 2014; GAO, 2008; Simpson, 2014.

A number of countries are making significant investments to improve the quality of mapping. For example in Australia, a National Work Program for Flood Mapping aimed at improving the quality and consistency of flood mapping has been established. Under the programme, an analysis of gaps in coverage of existing maps is being undertaken in order to support the prioritisation of future investments in flood mapping. To ensure national consistency, principles and technical standards for flood risk mapping are also being developed and its outputs will apply across all jurisdictions within Australia (and therefore, will aim to avoid differences in methodologies that might undermine the usefulness of flood maps for end-users, including the insurance industry). In the United States, FEMA requested USD 400 million annually over fiscal years 2013-2017 to develop more accurate flood maps and address some of the shortcoming identified in Box 4.7 (although FEMA has indicated that it will take several years to bring all the maps up-to-date (Simpson, 2014)). Turkey is also investing in flood mapping with the aim of completing risk maps by 2017.

In the European Union, the implementation of the Flood Directive (Directive 2007/60/EC) has been an important driver of improvements to mapping in many countries (Surminski et al., 2014). The Directive requires member countries to undertake detailed flood risk mapping for areas identified as high-risk (based on a preliminary flood risk assessment) and to update those maps every six years. For example, in Austria, the implementation of the Directive has required the completion of flood risk maps for 391 areas facing potentially significant flood risk. Ireland has completed a preliminary flood risk assessment and is nearing completion of detailed risk mapping for 300 communities facing potentially significant risk. Portugal has completed a flood risk and vulnerability mapping exercise that considers the potential impact of various climate change scenarios. The United Kingdom has completed national and locally-detailed surface water flooding maps in response to the Directive and recommendations from the Pitt Review (2008) on lessons from the severe 2007 flooding in England (Surminski et al., 2014). In Switzerland, cantons and communities are obliged by federal law to establish hazard maps for a series of natural perils, including floods, with a resolution down to the individual lot or parcel.

The availability of risk maps suitable for underwriting flood insurance coverage is often driven by demand from a private insurance sector in need of modelling and mapping in order to assess flood risk. In a US GAO study (2014) on means to enhance private sector involvement in providing flood insurance, those consulted anticipated private insurance company interest in underwriting flood risk would drive demand for modelling and attract risk modelling firms to invest in building accurate risk maps for US flooding. As the environment for private sector coverage of flood risk improves in the United States, new probabilistic inland flood models are already being released (Miranda, 2014). Similarly, in Canada, the announcement of the government's intention to "explore options for a national approach to residential flood insurance" (Department of Finance Canada, 2014) has coincided with the development of flood risk models for Canada by several major reinsurance brokers and catastrophe modelling firms.

Even where private sector modelling and mapping of flood risk is well-established, governments have a role to play in ensuring the availability of the data necessary for developing and maintaining flood risk models. Governments are a significant provider of satellite imagery and are also the major source of most meteorological and hydrological data for their territories. Investments in building the capacity of meteorological and hydrological services generally provide broad benefits (not just in terms of better data for modelling flood hazards). According to the World Meteorological Organisation, estimates of cost-benefit ratios for investments in meteorological and hydrological infrastructure are generally around 1:10 (Jarraud, 2007). The potential benefits of upgrading all developing country hydrometeorological information production and early warning capacities to developed country standards have been estimated to be between USD 4 billion and USD 36 billion on an annual basis (Hallegatte, 2012).

Technological advancements are making an important contribution to the quality of flood risk maps. The availability of higher-resolution digital terrain models have allowed for more granular flood risk modelling that is able to provide greater differentiation in risk levels based on more accurate assessment of the probability of inundation. Satellite imagery has provided modellers with an accurate picture of the footprint of past floods that can improve the calibration of existing models. The increasing availability of satellite data has also allowed greater geographical coverage of flood maps, including into areas where no other mapping or modelling capacity is available. Google Earth™ and other publicly-available satellite data allow underwriters to estimate critical variables such as the distance of a structure from water and the elevation of the structure relative to that body of water. These advancements could contribute to increased insurance availability for flood risk in countries where risk mapping (and even hazard mapping) is not available as insurers do underwrite commercial risks in many countries where risk maps are not available (Thistlethwaite and Feltmate, 2013).

While experience from past events should not alone drive the identification of areas at risk of flooding, data on past losses is critical for calibrating (and re-calibrating) flood models. Real-world experience provides modellers with information on key inputs such as levels of water absorption for a given precipitation event, levels of flood damage for a given floodwater level (and the accuracy of damage functions) and other variables. Catastrophe modelling firms will often release updates to models after significant (or unexpected) events, such as the 2013 flash flooding in Calgary and Toronto (Canada) (Boyle, 2015). However, among the (developed and developing) countries that responded to an OECD survey on the financial management of flood risk, only about one-third indicated that data on past insured flood losses was available. The limited availability of data on past losses has been cited as a limiting factor in the robustness of models in a number of European countries (Boyle, 2015).

In countries where detailed topographical data and catastrophe models do not exist, information on past events can provide an (imperfect) source of information for understanding flood exposure. Broad use of technology by those affected could potentially be used to "crowd-source" data on flood impacts in order to develop initial maps of potential flood exposure (MacClune et al. 2015).

As in the case of mitigation investments, there may be some benefit in coordinating investment in mapping with the needs of the insurance sector for risk information. The efforts to develop quality forward-looking flood maps in Portugal was partly driven by the need to address low insurance penetration rates by providing insurance companies with a basis for pricing flood risk. As noted, the prioritisation exercise for flood mapping in Australia also explicitly considers the needs of end-users of hazard maps, such as the insurance sector.

## 4.3 Addressing limited demand for flood insurance

As outlined in Chapter 3, a number of factors combine to reduce households' and businesses' willingness-to-pay for flood insurance coverage, including the tendency towards underestimation of risk, misunderstandings about coverage and expectations of post-disaster compensation or financial assistance. The following sections describe possible approaches to enhancing the willingness-to-pay for flood insurance, including efforts to enhance public awareness of flood risk and the need for financial protection against such risks as well as ways to reduce the negative impact of *ex post* government compensation on willingness-to-pay for insurance coverage. This is followed by a discussion of the roles of premium subsidies and various forms of compulsion and bundling of insurance coverage in addressing underinsurance of flood risk.

## *Enhancing public awareness*

### *Building understanding of risk levels*

There are a number of opportunities to enhance public awareness of the level of risk from flooding, ranging from making information on flood risk widely available to disclosure requirements related to property transfer or rental, to the use of price signals in setting insurance premiums. Almost all countries make information on flood risk publicly available. More than two-thirds of countries that responded to an OECD survey indicated that flood hazard maps are made publicly available and that various types of awareness campaigns are implemented to enhance understanding of flood risk. For example, in Japan, the Flood Control Act requires municipalities to distribute relevant information on flood risk to residents. In the United States, the FloodSmart program provides information on flooding and flood risk. In the United Kingdom, the "What's in your backyard" application allows users to access flood risk (and other environmental) information at the post code level. In France, a major public awareness campaign surrounded a major flood exercise for a 1-in-100 year flood affecting the Paris region, including videos of the potential impacts of flooding on various Paris landmarks. In Australia, a National Flood Risk Information Project has been implemented to improve the quality, availability and accessibility of flood risk information, flood hazard data and flood-related imagery (including from past events). However, concerns about liability related to the accuracy of flood risk information have led to some reluctance to publishing all relevant information (similar to the liability concerns related to land-use planning, see Box 4.2) (The Australian Government the Treasury, 2011).

A significant opportunity to build awareness arises at the time of purchase or rental of a residence and, in some countries, information on flood (and other hazards) at the level of individual properties is communicated at that time. For example, in France, sellers and landlords are required to provide information on any compensation that has been paid in relation to the property as a result of a natural (or technological) disaster and the risk of flooding must be disclosed as part of the home purchase process. Such information is also available in Australia (as "vendor statements") with some states (e.g. Victoria) deemed to be providing a robust system of disclosure of flood and other natural hazard risks (Productivity Commission, 2014). In other countries, property-level risk information is available but is not automatically disclosed upon property transfer or rental. In New Zealand, a Land Information Memoranda that provides information on natural hazard risks associated with a property or structure is available from the local council to any party upon request and payment of a fee. In England and Wales, the Environment Agency and Natural Resources Wales can provide households with a detailed flood map or a letter setting out the flood risk from rivers and the sea for the area of their property which can be used for the purposes of securing insurance coverage (DEFRA, 2013).

There is some evidence that public awareness levels will be affected by the approach taken to communicating risk. For example, a 1-in-5 chance of a flood over 25 years has been demonstrated to be taken more seriously than a 1-in-100 chance of a flood on an annual basis, even though the two describe similar levels of risk (0.8% annual probability vs. 1.0% annual probability) (Kunreuther and Weber, 2014). The use of return period probability measures (i.e. 1 in a given number of years) may give some the mistaken impression after an event that they will be safe from flooding for the remainder of the return period (GAO, 2014). There is also a general tendency to assume that the actual occurrence of the event will occur near the end of the return period which suggests that using return periods within the lifespan (or normal property ownership period) of an individual homeowner may be more effective (Henrich, McClure and Crozier, 2015). The use of descriptions of impact, such as potential number of deaths, amount of damage, etc., can also improve the effectiveness of risk communication as there is some evidence that homeowners also tend to underestimate the potential impact of a flood. For example, a survey of New York City residents found that only about one-third of respondents had a relatively accurate perception of their probable flood damage and that more people tend to underestimate than overestimate the level of flood damage that they could face (47% underestimate and 19% overestimate) (Botzen, Kunreuther and Michel-Kerjan, 2015).

Experience with flooding has also been shown to play a significant role in decisions to purchase insurance coverage (or implement risk reduction measures) (see Box 4.8) which suggests that there might be some advantages to using risk communication approaches that are based on past events.

Box 4.8. **The benefit of flood experience for risk reduction and financial protection**

In many countries, the actual experience of being impacted by a flood event has often been a significant driver of the demand for flood insurance (while a lack of experience with an event may lead households to not renew their coverage). A systematic analysis of flood insurance take-up rates in the United States found an increase in take-up rates of 9% in areas affected by a presidentially-declared (flood) disaster, followed by a return to "normal" levels over a period of approximately 9 years (Gallagher, 2014). Similarly, a study of homeowners on the US Atlantic Coast found that those that had experienced previous hurricane damage were more likely to seek hurricane and flood insurance (Hudson et al., 2014). Substantial increases in insurance coverage for flood risk have also occurred after flooding in Germany and Australia (Swiss Re, 2012) and the 2005 hurricane season in the United States (Michel-Kerjan, 2010).

This suggests that (large) flood events may offer an opportunity for enhancing the effectiveness of public awareness campaigns. In Germany, the experience of local flooding in 2000 combined with a dedicated advertising campaign was effective in increasing flood insurance take-up rates more broadly. The study of post-event take-up rates in the United States also found an increase in take-up rates of 3% in neighbouring communities that weren't directly affected by flooding and in communities that were part of the same media market and therefore exposed to similar information on the floods (even where the communities with a shared media market were geographically distant and dissimilar to the affected community) (Gallagher, 2014).

*Source*: Gallagher, 2014; Hudson et al., 2014; Swiss Re, 2012; Michel-Kerjan, 2010.

A key challenge is communicating risk to households and businesses facing relatively lower-risk of flooding. For example, the demarcation of flood zones with more frequent return periods (e.g. 1-in-100 year SFHAs in the United States) may give the impression to those outside such zones that they face no flood risk at all. Similarly, the construction of flood protection barriers may give those within the zone of protection an unwarranted sense of being completely protected against any future flood risk.

Insurance premiums that are risk-based can also offer an important signal on the level of risk faced by individual households or businesses. Countries with flat-rate pricing of insurance premiums or public (re)insurance backing for high-risk properties (where invisible to the policyholder) do not communicate risk levels to policyholders (Surminski and Eldridge, 2014). Risk-based premiums are also important for ensuring that property-level risk information is transmitted through property values. A property facing significant flood risk and high insurance premiums should also have a lower resale value (all else equal). While this is difficult to measure (given the multitude of factors that affect house prices), a study in the United States found some evidence of price differentials for properties due to different levels of flood risk, particularly in periods after the occurrence of an event in the given community (with a decline in that differential over time) (Bin and Landry, 2013). Mandatory property-level risk disclosure could be a means of maintaining risk awareness in the years following an event. For example, a study on the impact of the 1998 California Natural Hazard Disclosure Law, which requires sellers of properties to disclose whether properties are located in SFHAs (among other natural hazard zones), found evidence that the law created a price differential of -4.2% for houses located in an SFHA (Troy and Romm, 2004).

### *Improving understanding of financial protection*

In addition to enhancing household and business understanding of flood risk, many countries also invest in enhancing awareness of the need for financial protection. Just under one-half of surveyed countries indicated that information on financial protection options, such as insurance, are included in public awareness initiatives related to flood risk. In some countries (e.g. Czech Republic), a lack of financial education related to the protection that insurance can provide is seen as a significant cause of underinsurance given that awareness of flood risk (especially as a result of recent experience) is high. In the United Kingdom, a guide to obtaining flood insurance in high-risk areas was developed in collaboration with the National Flood Forum, the insurance industry and others (DEFRA, 2012). In the United States, FEMA's FloodSmart program provides information to the public on the benefits of purchasing flood insurance.

Where flood coverage is an optional add-on, there may be misperceptions among consumers about the level of flood insurance coverage included in their home, commercial, contents and/or business interruption policies. In Italy, for example, 42% of respondents to a survey believed (wrongly) that they were insured against damage from natural catastrophes (Swiss Re, 2013). In Australia, a survey of homeowners found that 37% were unsure as to whether there building insurance covered damage from flooding, including 23% of those that lived in known flood risk zones (Quantum Market Research, 2013). Based on the experience of the 2011 Queensland flooding, the Australian government has imposed new regulatory requirements to ensure that a standard definition of 'flood' is used in home building, home contents, small business and strata title (i.e. commonly-owned areas in residential buildings) insurance policies, and to require insurers to provide consumers with one-page fact sheets that set out key information about the coverage provided under home building and home contents insurance policies (OECD, 2015a).

How an optional flood coverage is offered to policyholders (i.e. opt-in vs. opt-out) can also have important implications for the level of take-up of flood coverage. Individuals tend to have a bias towards the default option when offered different options and there are a number of examples from other fields where differences in the default option offered result in differences in take-up rates (e.g. rates of organ donation in countries where organ donation is automatic upon death are much higher than in countries where organ donation requires *ex-ante* consent (Johnson and Goldstein, 2003) as are rates of enrolment in defined contribution pension plans that use auto-enrolment (OECD, 2013b)). In Japan, the insurance industry instituted such a practice by specifically confirming a customer's decision not to extend their coverage to include protection against earthquakes as a means of reducing the potential for misunderstandings of the level of financial protection (Orie and Stahel, 2013).

The automatic renewal of insurance coverage (as in Germany) or the offering of flood insurance coverage as a multi-year contract could have a similar effect by making the maintenance of coverage over a number of years the default option (National Research Council, 2015). A survey in the Netherlands found that individuals might even have a higher willingness-to-pay for multi-year contracts relative to annual contracts (Botzen, de Boer and Terpstra, 2013) (although such contracts will usually be more costly than an annual contract due to uncertainty related to potential changes in risk or reinsurance market conditions and are prohibited in some countries, such as Germany, for contracts longer than three years).

In addition to improving the understanding of financial protection (and facilitating its purchase), some countries identified a specific need to raise the public's awareness of their responsibility for protecting themselves against flood risk. The expectation of government assistance can reduce the incentive for seeking financial protection as a costless alternative to the protection provided through insurance coverage. This impact can be somewhat mitigated by limiting the amount of financial assistance to small amounts or by only providing compensation for losses that are truly uninsurable. Lump-sum payments made irrespective of insurance coverage would also limit the impact of financial assistance on demand for insurance coverage (Schwarze et al., 2011).

However, even where the amount of government financial assistance available is limited, the perception that large amounts of government funding might be available could lead to misunderstandings about the need for financial protection. The publicity surrounding major flood events and the substantial public funding dedicated to recovery and reconstruction could give the impression that significant funding is available to affected households or businesses, when in reality most of the funding will usually be allocated to the reconstruction of public infrastructure. High ceilings on potential financial assistance, even when rarely reached, can also exacerbate misperceptions about access to public compensation. In the United States, FEMA's Individual Assistance program can provide up to USD 31 500 per household although the average grant size is USD 4 000 (Kunreuther and Michel-Kerjan, 2013) (large amounts are also available to households and businesses from the US Small Business Administration although that assistance is provided only as a loan (GAO, 2014)).

In most countries, the level of post-disaster government assistance that might be provided to an affected household or individual depends on a number of factors and is difficult to estimate in advance. In Germany, for example, government assistance is not based on formal legislation and therefore the granting of such assistance may not happen by default (Thieken et al., 2006). In the Netherlands, a statutory compensation mechanism has been established but there are no predefined rules on eligibility for – or level of – compensation (Surminski et al., 2014). In Russia, the government is legally obligated to provide assistance after a disaster although the amount of such assistance is only determined after the event. In many countries, government assistance is only available where there is a government declaration of a disaster which may not occur in the case of smaller floods. Even in countries where specific programmes and eligibility criteria have been established for post-disaster assistance, the implementation of those programmes may be inconsistent. For example, the Australian Productivity Commission inquiry on natural disaster funding arrangements found instances of inconsistent application of the Australian Government Disaster Recovery Payment based on different applications of the eligibility criteria across different disasters (Productivity Commission, 2014).

There may be advantages in attempting to provide greater clarity on the amount of assistance likely available post-disaster as well as the conditions for accessing such assistance as a means of addressing these types of misperceptions (although this will create moral hazard if the assistance is not limited to uninsurable losses). A financial decision support tool that compares the compensation that would be available through insurance relative to government assistance could provide households with a means to more accurately assess their financial protection options (National Research Council, 2015).

### *Premium Subsidies*

Enhancing understanding of flood risk and financial protection may not be sufficient in increasing coverage where premiums are unaffordable. In such cases, the provision of subsidies may be an option for ensuring that households (and potentially, businesses) are protected against flood risk – although, as outlined below, the use of premium subsidies can be expensive, difficult to remove, are likely to exacerbate moral hazard and have limited (or no) impact in terms of reducing the level of risk.

Premium subsidies can be effective in terms of increasing coverage. While demand for flood insurance coverage has been found to generally be price inelastic (National Research Council, 2015), studies have found a positive correlation between income and the amount of insurance purchased. In the United States, for example, income was found to be positively correlated to the amount of flood insurance purchased (Browne and Hoyt, 2000), while the purchase of optional insurance for contents in the United Kingdom has been found to be positively correlated with income (DEFRA, 2013). In a survey of insurance customer preferences across 30 countries, value for money was deemed to be the most important factor in decisions to purchase non-life insurance coverage while cost was the factor most often-cited for ending non-life insurance coverage or switching providers (Ernst & Young, 2014).

A number of countries provide subsidies to support the affordability of flood insurance, including both direct premium subsidies (explicit subsidies) for high-risk properties, cross-subsidies (or implicit subsidies) resulting from pricing that is not completely risk-based and tax exemptions related to the payment of premiums, either for consumers as expenses or insurers as revenues. In some countries, cross-subsidies have been the result of insurance sector practices, not specific government policy. In the United Kingdom, for example, until recently, insurers have generally (and voluntarily) charged similar amounts for home insurance with flood coverage to households facing very different levels of flood risk (DEFRA, 2013). In Japan, premiums for household coverage that includes flood are based on prefecture and building structure, regardless of the level of flood risk. In Chile, semi-flat premiums are charged for flood insurance coverage due to limitations in modelling that impede the establishment of true risk-based premiums. In other countries (France, Spain, Switzerland (for risks covered by the private sector)), cross-subsidies resulting from flat (or relatively flat)[1] pricing for flood (and other natural disaster) coverage is a deliberate government policy based on a principle of solidarity and sharing of risk across citizens.

Premium subsidies targeted to high-risk properties are provided in the United States and in the United Kingdom. In the United States, premium subsidies are provided for structures constructed before the completion of a Flood Insurance Rate Map (FIRM) for the given area (known as pre-FIRM structures) and for structures that have been re-zoned into a higher-risk area based on the completion of a new FIRM (grandfathered structures). The subsidies apply to the first USD 35 000 of coverage, with coverage for amounts above that threshold priced at full risk rates. On average, the property owners that benefit from such subsidies pay 35-40% of the full-risk rate for flood coverage (CBO, 2007). The subsidies (pre-FIRM and grandfathered) remain available if the property is sold although not after the property sustains substantial damage (a loss equivalent to 50% of the structure's market value) or if the property benefits from substantial improvement (leading to an increase of 50% or more in the structure's market value) (CBO, 2007). The subsidies are provided in the form of reduced premiums paid to the NFIP. Discounted rates are also available for properties outside of SFHAs with favourable loss histories (preferred risk policy) and for properties in communities that participate in Community Rating System (see Box 4.3). In the United Kingdom, premium subsidies are provided indirectly through the pricing of reinsurance for the flood portion of bundled household policies that is available through Flood Re. Flood Re offers standard prices for reinsurance coverage based on council tax valuation bands, no matter the level of flood risk. Insurers are free to set the premiums for the bundled coverage although the ability to transfer the flood risk component to Flood Re at a set price provides a notional ceiling on the premium rates for the flood component of household insurance coverage for high-risk properties (DEFRA, 2013; Flood Re, 2015).

Premium subsidies are not costless and can create significant contingent liabilities for public insurance schemes. In the United States, the cost of premiums subsidies is reflected in the significant losses that have been incurred by the NFIP after extreme events such as Hurricane Katrina in 2005 and Hurricane Sandy in 2012 (covered through borrowing from the US Treasury – see Chapter 5). These losses were incurred because the lower premium income resulting from subsidised coverage did not provide sufficient reserves to cover claims. In the United Kingdom, the costs have been incurred by low-risk policyholders that provided an estimated GBP 150 million in annual benefits to the approximately 250 000 households facing significant flood risk by paying higher premiums than their level of flood risk would require (DEFRA, 2013). Under Flood Re, the subsidy costs will be borne by the insurance industry (and their customers). Cross-subsidies, on the other hand, will not lead to direct costs if the overall level of premium income is sufficient to meet the cost of claims.

Premium subsidies (both those targeted to specific policyholders and cross-subsidies resulting from flat pricing) dampen the risk signal inherent in risk-based premium pricing and may therefore reduce (or eliminate) the policyholders' incentive for making investments in risk reduction. To avoid misperceptions about the level of risk as a result of such subsidies, it is important to ensure that policyholders are aware that the premiums they pay do not reflect the full level of risk that they face from flooding. In the United Kingdom, Flood Re has made an agreement with cedant insurers to pass information on the level of flood risk to their customers that benefit from Flood Re coverage. Where subsidies eliminate or reduce impediments to insurance affordability and availability, they may also reduce the incentives for mitigation or land-use controls at the level of local administrations (Douglas, Bowditch and Ni, 2013) (although this can potentially be mitigated by effective restrictions on land-use in flood zones and robust building codes).

A key consideration (beyond cost) is the impact of the subsidies on overall levels of financial protection. As noted, there is some evidence that subsidies are effective in increasing the demand for flood insurance. A study of demand for flood insurance among coastal households in the United States found evidence of higher demand for flood insurance coverage (and greater price elasticity of demand) from households that benefit from explicit premium subsidies (Landry and Jahan-Parvar, 2011). The proposal to eliminate subsidies under the *Biggert–Waters Flood Insurance Reform Act of 2012* coincided with a decline in policies in force in most US states, particularly among discounted pre-FIRM policies whose rates had risen with the implementation of the Act (Kousky and Shabman, 2015). Cross-subsidies that lead to higher premiums for low-risk policyholders could lead to reduced demand for coverage among low-risk households (Douglas, Bowditch and Ni, 2013) although that will likely depend on the cost per household of the cross-subsidies as well as the level of awareness that such cross-subsidies are being provided. In large policyholder communities, the cost per household would likely be low. For example, in the United Kingdom, the industry has estimated the cross-subsidy to be GBP 8-9 per year, per low-risk household on building and contents coverage (DEFRA, 2013).

The design of any premium subsidy programme will be important for controlling costs and ensuring that incentives for risk reduction are maintained (to the extent possible). Some considerations include:

- Subsidies should be means-tested to ensure that the benefits accrue to those most in need. To achieve this, a definition of affordability could be established with only those facing premiums beyond the affordability threshold benefiting from the programme. Such a threshold could be based on premium as a share of income or property value or some measure based on overall housing costs (including flood premiums) as a share of income (National Research Council, 2013).
- The scope of the programme could be limited to residential property only or only to those that are required to purchase insurance coverage (for example, as a condition of their mortgage financing). Neither the United States nor United Kingdom provide subsidies for high-risk commercial property.
- Costs could also be reduced by designing the programme to incur the cost of the subsidy upon the occurrence of the event (Allianz Australia Insurance Ltd., 2011), rather than at the time of the purchase of the policy. This is the case in the United States and United Kingdom as well as France, Spain and Switzerland as any costs related to subsidisation of premiums are only incurred when premium income is insufficient to meet claims. Another approach could be to provide support for the policyholder deductible, which would allow policyholders to choose higher deductibles (and therefore reduced premiums) and incur costs only upon the occurrence of an event. However, both of these approaches lead to uncertain contingent liabilities for government that will need to be appropriately managed (see Chapter 5).
- Another potential means to contain the costs of subsidies would be to limit the benefit of any subsidy to existing property owners (i.e. no subsidy would be available if ownership of the property is transferred). This could potentially impact the sale value of properties that benefit from the subsidy (although this could be seen as an appropriate correction to prices inflated by artificially low insurance premiums). Consideration should also be given to limiting subsidies to existing properties (i.e. not new developments) which recognises that some communities were built before the true level

of flood risk was known while ensuring that the availability of subsidies does not provide an incentive for developing risky areas.

- Investments in risk reduction could be required as a prerequisite to receiving the premium subsidy to counteract the impact of subsidies on a property owner's incentive to reduce their exposure. Similar to communities, financing for such investments should be available based on the expected reduction in insurance premiums that result (and could be provided through a public sector loan). Means-tested vouchers could also be provided to households that make risk reduction investments tied to the expected savings in risk-based premiums that will result from such investments (Kousky and Kunreuther, 2014). Mitigation loans could be attached to the property/mortgage (rather than the owner) to avoid complications related to ownership transfer (Knowles and Kunreuther, 2014).

- The premium subsidies could be funded (in full or in part) by local governments in order to provide an incentive for local communities (that usually make decisions on land-use and protective infrastructure) to reduce risk at the community-level (Allianz Australia Insurance Ltd., 2011; National Research Council, 2015). However, this would be less effective in older communities with legacy exposures to flood risk that are difficult to reduce.

In countries where national or local taxes are imposed on premiums for flood insurance (or insurance coverage more generally), consideration should be given to whether the imposition of such taxes impacts insurance affordability and is worth maintaining for all (or subsidised) properties. In Australia, for example, premium taxes have been estimated to increase premiums by 44% for home insurance and, based on modelling, may be responsible for close to 70 000 households choosing not to secure home insurance coverage (Barker and Tooth, 2008). Allowing insurers to accumulate tax deductible reserves could also potentially increase affordability by reducing the cost of insurance coverage.

The establishment of a public insurance scheme itself is a means of subsidising or supporting premium costs, depending on the structure of the scheme and rates charged. Public schemes can promote affordability by charging lower than actuarially-based premiums and also by passing on any cost-savings resulting from their status as a public entity, such as lower financing costs, limited marketing costs, tax savings as a result of income tax and/or premium tax exemptions as well as the underwriting profits and return on capital normally required by a private insurer. The scope of these cost savings could be significant. For example, in the United States, income taxes and the cost of capital have been estimated to account for approximately 20% of premiums (Property Casualty Insurers Association of America, 2011). In Australia, taxes and underwriting profits have been estimated to account for 17-33% of premiums (Douglas, Bowditch and Ni, 2013). A study of insurance premiums in three regions of Switzerland, Austria and Germany facing similar levels of flood risk found that the monopoly public insurer in Switzerland was able to offer premiums at rates of about one-third of premiums charged in the other regions (although insurance is also mandatory in that region of Switzerland which allows for the establishment of a broader risk pool) (Schwarze et al., 2011). However, these costs savings need to be considered in the context of any increase in operational costs involved in operating a public scheme relative to the operating cost of private insurers (see Chapter 5) as well as the costs to governments in terms of foregone tax revenue (where a public insurer is tax exempt).

Once provided, subsidies are extremely difficult to eliminate. In the United Kingdom, the elimination of cross-subsidies would lead to an estimated 200 000 households facing increases in premiums of 2% or more of household income (Flood Re aims to eliminate subsidies over 20-25 years) (DEFRA, 2013). In the United States, the *Biggert-Waters Flood Insurance Reform Act of 2012* reformed the NFIP to eliminate, over time, subsidised rates for certain policies, which would have led to significant long-term premium increases in a number of communities. For example, flood insurance rates for some properties in Hawaii, Georgia, Louisiana, and other states were reportedly set to increase from USD 600 annually to USD 20 000 - 50 000 (Nance, 2015). However, before these reforms could be fully implemented, the *Homeowner Flood Insurance Affordability Act of 2014* was enacted, which delayed, and in some cases eliminated, many of the rate reforms under the *Biggert-Waters Flood Insurance Reform Act of 2012* (see Box 4.9).

### Box 4.9. **Reform of premium subsidies in the United States**

In an effort to address some of the challenges related to the NFIP, including the significant deficits created by Hurricanes Katrina and Sandy, the US Congress passed the *Biggert–Waters Flood Insurance Reform Act of 2012*. The Act required the elimination of subsidised flood insurance premiums over a transition phase through:

- the phase-out of all subsidised premiums for businesses and non-primary residences over 4 years;
- the phase-out of all subsidised premiums for all repetitive loss properties over 4 years;
- the phase-out of all subsidised premiums for "grandfathered" properties over 5 years (i.e. properties that were re-zoned into a higher-risk area based on the completion of a new FIRM);
- the elimination of access to subsidised premiums for any new purchases, new policies, property transfers, policy lapses or new repetitive loss; and
- the phase-out of all remaining subsidised premiums, including primary residences, beginning in 2014.

In response to the new requirements, a Coalition for Sustainable Flood Insurance was formed with the support of communities from 35 US states which lobbied the federal government to reverse the requirements. In 2014, the US Congress passed the *Homeowner Flood Insurance Affordability Act of 2014* which delays – but does not halt - the implementation of the *Biggert-Waters Flood Insurance Reform Act of 2012* by repealing some clauses and establishing new subsidies for some types of policies. The *Homeowner Flood Insurance Affordability Act of 2014* reinstated premium subsidies for grandfathered properties (i.e. properties that have been re-zoned into an SFHA) and will continue to allow property owners re-zoned into SFHAs in the future to have their rates grandfathered. Premium subsidies for pre-FIRM structures, on the other hand, will continue to be phased out (National Academies of Sciences, Engineering, and Medicine, 2015).

*Source*: National Academies of Sciences, Engineering, and Medicine, 2015.

The opportunity cost of premium subsidies, which are recurring and have limited (or no) benefit in terms of reducing risk, needs to be evaluated against the alternative use of those funds for investing in risk reduction. As noted above, investments in property-level risk reduction such as property elevation, while capital intensive up-front, can lead to reductions in premiums that more than offset the cost of financing the initial investment. Consideration should be given to whether subsidies for risk reduction would be a more efficient use of public funds than premium subsidies. The establishment of programs to encourage household risk reduction (information on risk reduction options as well as grants and loans) could be one means to facilitate a transition away from subsidised premiums (Association of Floodplain Managers, Inc., 2013).

### *Compulsion*

Other approaches to ensuring sufficient financial protection among households and/or businesses relate to making insurance coverage against flood risk mandatory or tying requirements related to insurance coverage for flood risk to mortgages, government compensation or other assistance or to home insurance coverage more generally (i.e. automatic extension of home insurance policies to include flood risk or bundling of flood insurance with home insurance).

There are three main types of insurance compulsion: mandatory purchase, automatic extension and mandatory offer (see Table 4.1).

There are a few countries that impose mandatory insurance requirements. In Iceland and most (22 of 26) Swiss cantons, insurance coverage of flood (and other disaster risks) is mandatory for all residential and commercial buildings. These requirements have been successful in ensuring broad financial protection against disaster risks as penetration rates are effectively 100%. However, in both cases, the requirements are complemented by extensive public intervention in the private insurance market through provision of direct insurance by the public sector and/or pricing regulation. In Iceland, the insurance coverage is provided by a public entity, Iceland Catastrophe Insurance (ICI), at a flat rate. In 19 of 23 Swiss cantons, (mandatory) insurance coverage for buildings is provided by a public cantonal monopole insurer with some limited risk pricing. In the other six cantons, the private sector provides the coverage for disaster risks although the pricing for that coverage is regulated and completely flat. Price regulation is necessary to eliminate opportunities for private companies to seek high profits in a market with mandatory demand. A pool has also been established by the private insurers to distribute disaster losses across insurers according to their market share in order to prevent the private insurers from only offering coverage for low-risk structures.

A mandatory insurance requirement with insurance provided by the public sector is economically similar to *ex post* government compensation for all losses (as in both cases, the costs are ultimately borne by taxpayers), although with some important differences. Public insurance offers the advantage of pre-funding those losses through the collection of premiums. While a dedicated general government reserve fund could be established, it would likely be more susceptible to being diverted to other uses (relative to the reserve fund of a public insurer). In some countries (e.g. France, United States), the accumulated reserves of the public insurer are also put to good use as a source of funding for prevention. A public insurer could also vary premiums based on risk and therefore provide incentives for risk reduction (Schwarze et al., 2011). On the other hand, where pricing is relatively flat, the overall cost for mandatory insurance will vary with the value of the structure/property which may be less equitable than *ex post* government compensation of losses that would be distributed across taxpayers based on income.

Table 4.1. **Types of insurance compulsion**

| | Advantages | Disadvantages |
|---|---|---|
| Mandatory purchase | Promotes the expansion of disaster insurance coverage, and the diversity of risks covered, which should help to reduce insurance costs overall.<br>Eliminates the risk of adverse selection (i.e. those who perceive themselves to not be at risk may not purchase insurance, possibly increasing risks in the pool).<br>Addresses potential behavioural biases, which may otherwise lead to inadequate coverage.<br>Serves to clarify the allocation of disaster costs and reduces the government's implicit contingent liabilities. | May be unpopular and perceived as a tax.<br>May run contrary to the culture of the country and face constitutional constraints (e.g., limit to private autonomy).<br>Enforcement of purchases may be difficult.<br>Given the captive market, the insurance sector may seek to build strong profit margins into premium rates; at the other extreme, inadequate pricing may lead to underwriting losses and drain capital from the industry.<br>Mandated pricing may become overly influenced by other policy objectives. |
| Automatic extension<br>i.e., automatic inclusion of disaster coverage in basic voluntary property insurance policies | Can be effective if the penetration rate of the underlying basic policies is relatively high, so that they are used as a vehicle to spread disaster insurance coverage.<br>Compared with the mandatory purchase of disaster insurance, this option entails a lower degree of compulsion and may be less unpopular. | May have negative effects on the market for the basic property policy to which the mandatory disaster extension applies if the extension leads to higher premiums or the exit of some insurers that are unwilling to underwrite disaster risk. |
| Mandatory offer | Promotes the expansion of disaster insurance coverage, so that businesses and individuals who are willing to purchase financia protection can do so. | May lead to adverse selection: those who perceive themselves to not be at risk may not purchase insurance, possibly increasing risks in the pool and leading to sub-optimal take-up rates; low risk awareness or cognitive biases may aggravate this effect. If the penetration rate remains very low, there may be inadequate risk pooling. |

*Source*: Adapted from OECD (2012).

Such an approach may not be feasible in all countries. There may be public resistance to mandatory insurance requirements which could be perceived as a tax. There also may be constitutional impediments to mandatory insurance requirements in some countries. For example, in Germany, an infringement on personal autonomy, such as the imposition of an insurance requirement, is only permissible where that infringement is in the public interest and is appropriate and proportionate (i.e. there is no less intrusive way to achieve the same outcome) (Schwarze and Wagner, 2007).

Mandatory automatic extension of insurance policies to include coverage of disaster risks is more common. In France, insurance coverage for disaster risks is provided based on a mandatory surcharge applied to all home, commercial (including business interruption) and motor vehicle insurance policies. Insurers may choose to maintain the full disaster exposure covered by the surcharge or transfer it to a public (re)insurer or private reinsurance markets with the transfer of risk to the public reinsurer (*Caisse centrale de reassurance (CCR*) limited to 50% of the exposure. In Spain, coverage for extraordinary risks is mandatorily included in property, life and personal accident policies (and subject to a mandatory surcharge) and may be transferred in full to the *Consorcio de Compensación de Seguros* (CCS), a public entity, or retained in full by the private company. In Belgium, mandatory extension of insurance coverage to most disaster risks is also a requirement. These approaches have been successful in achieving high rates of penetration of flood coverage (approximately 75% in Spain, above 75% in Belgium, and 97% in France, for residential properties) although they are dependent on high-levels of property insurance coverage (upon which the extension is added). They also benefit from the creation of a broad pool of low and high-risk properties as well as diversification across different perils.

In countries where flood risk is not generally covered at present, mandating a new requirement could lead some companies to exit the home insurance market altogether rather than attempt to build capacity for underwriting flood risk. It could also lead to reduced competition in standard property insurance markets for properties in high flood risk areas as insurers may choose not to provide any property coverage in these areas if they are also forced to provide coverage for flood risk. While there would be expected benefits to the private sector from mandatory insurance requirements (in terms of creating/stimulating demand), there could also be resistance if the private insurers are forced to provide coverage for flood risks for all properties (including those considered uninsurable). Mandatory extension of insurance policies to cover disaster risks, where it leads to higher premiums, may also affect the demand for home insurance policies more generally. In Belgium, a public institution, *Bureau de tarification – Catastrophes naturelles*, has been established to arrange premium rates and contractual conditions for natural disaster risk policies that insurers refuse to cover under their own terms (or where the premiums that private insurers offer are unaffordable) (OECD, 2015a).

In a number of countries, including Austria, Latvia, Poland, Russia and Turkey, insurance coverage for flood risk is an automatic (although not mandatory) extension to most home insurance policies. In Austria, a limited amount of flood coverage is provided on a first-loss basis as part of standard home policies – additional coverage is available as an add-on. In Latvia, Poland, Russia, and Turkey, insurers are able to exclude flood damage from these policies where they deem the risk to be too high. In Latvia, the automatic extension of insurance policies to cover flood risk has achieved very high levels of penetration (95%) although levels of penetration in Austria, Russia and Turkey are much lower (less than 25%). Mandatory offer of flood coverage can also be beneficial in terms of ensuring awareness of flood risk and financial protection options, and can make an important contribution to increasing overall coverage levels (particularly where the offer is made on an opt-in basis, as discussed above). In the United Kingdom, the Gentleman's Agreement and Flood Re involve a commitment by insurers to offer flood coverage to all properties (built prior to 2009, as outlined above).

Where property insurance coverage is automatically extended to cover both wind and flood damage (for example, from a cyclone), the need for assessing the cause of damage – which can be challenging where structures are completely swept away – would be eliminated, allowing for more efficient claims settlement. For example, Hurricane Katrina led to years of legal disputes over whether homes were taken away by wind or water (known as "slab lawsuits") (Sandink et al., 2010). This is the approach that is taken in France and Spain (where the mandatory extension of property insurance covers a number of disaster perils) and the approach proposed in a study on the establishment of a reinsurance pool to address insurance affordability in Northern Australia (The Australian Government the Treasury, 2015b).

In a number of countries, mortgage lenders will generally require flood coverage on the assets against which they are providing financing. In the Czech Republic and Portugal, this practice is perceived as a major driver of flood insurance penetration. In the United Kingdom, a requirement that flood insurance coverage be maintained for the duration of the mortgage in order to comply with the terms of the loan has been seen as key in achieving penetration rates of over 90% (DEFRA, 2013).

In the United States, there has been a statutory requirement for federally-regulated mortgage lenders to ensure that flood insurance coverage is in place for homes in SFHAs to which they lend. The requirements were tightened in 1994 after large floods on the Missouri and Mississippi rivers demonstrated the extent of continued underinsurance for flood risk (less than 20% of the flooded structures were insured) (Galloway, 1995). The 1994 changes included requirements that: i) coverage be maintained over the life of the loan; ii) flood insurance payments be escrowed where an escrow is required; iii) lenders obtain flood insurance coverage where borrowers do not (a recent proposal would clarify that lenders can charge homeowners for that insurance); and iv) failure to comply with the requirements could lead to fines imposed against lenders (National Research Council, 2015). However, an analysis of the NFIP insurance portfolio over 2000-2009 found that flood insurance policies tended to lapse after 2-4 years, including in SFHAs, despite the requirement for maintaining coverage over the life of the loan. As the average length of residence (7 years) is much longer than the average policy length (Michel-Kerjan, 2010), the causes likely include a general failure by lenders to verify coverage except at origination, the practice of transferring mortgages to other financial institutions or capital markets where verification of compliance with the flood insurance requirement would be unlikely to occur as well as limited fines imposed for non-compliance (Kunreuther and Michel-Kerjan, 2013). Lenders are also not required to monitor any change in the flood risk affecting a mortgaged property, for example, should a remapping lead to the inclusion of a property inside an SFHA (Huber, 2012). Beyond the enforcement challenges, the effectiveness of mortgage-linked requirements would also be limited in regions where mortgages on properties are less common. For example, an analysis in the United States of nine coastal counties found that only 39% of beach houses and 34% of properties located in the flood zone had mortgages (Landry and Jahan-Parvar, 2011),

In the United States, eligibility for public disaster assistance is also tied to insurance. For households in SFHAs, their communities must participate in the NFIP in order for them to be eligible for federal disaster assistance programs. Individual households that access these programs must also purchase flood insurance as a condition for receiving assistance. In addition, in South Carolina, policyholders seeking coverage through the state-backed wind pool are required to have federal flood insurance (Hartwig and Wilkinson, 2014).

In the United States, another form of compulsion has been proposed in the form of community-based flood insurance (a study of this option was mandated under the *Homeowner Flood Insurance Affordability Act of 2014*). A community-based flood insurance option would involve issuing insurance coverage at the community-level with premiums funded by community members through a specific charge or through tax revenues. This approach has a number of potential advantages, including the establishment of a large pool of diversified risks (depending on the size of the community involved) and the creation of incentives for risk reduction at the community-level (where responsibility for land-use planning and flood protection investments often lies) through the possibility of lower premiums.

## Note

1. In Switzerland, the premium rates charged by cantonal monopole insurers are partially cross-subsidising, although some structures may face premiums of three times the base rate or even up to a factor of 100 in rare cases (e.g. in one of the cantons: glasshouses and buildings with an extraordinarily bad loss experience may face premium rates of 40 per mill instead of 0.4 per mill of the sum insured).

## References

Adaptation Sub-Committee Secretariat (2014), *Policy note: flood and coastal erosion risk management spending*, Committee on Climate Change, London, 21 January, *www.theccc.org.uk/publication/policy-note-flood-and-coastal-erosion-risk-management-spending/,* accessed 12 January 2016.

Aerts, J. C. J. H. et al. (2008), "Dealing with uncertainty in flood management through diversification", *Ecology and Society*, Vol. 13(1).

AIR Worldwide (2015), *Katrina 2015*, 26 August, *www.air-worldwide.com/Publications/AIR-Currents/2015/Katrina-2015/*, accessed 11 January 2016.

Allianz Australia Insurance Ltd. (2011), *Natural Disaster Insurance Review Submission*, 21 July.

Association of British Insurers (2009), *Climate Adaptation: Guidance on Insurance Issues for New Developments*, Association of British Insurers, London.

Association of State Floodplain Mangers, Inc. (2013), *Flood Insurance Affordability: ASFPM Recommendations to address the impact of NFIP Reform 2012 (BW-12)*, Association of State Floodplain Mangers, Inc., Madison, WI.

The Australian Government the Treasury (2015a), *Northern Australia Insurance Premiums Taskforce: Interim Report*, Commonwealth of Australia, *www.treasury.gov.au/~/media/Treasury/Consultations%20and%20Reviews/Reviews%20and%20Inquiries/2015/NAIP%20Taskforce/Interim%20Report/Downloads/PDF/NAIP-interim-report.ashx.*

The Australian Government the Treasury (2015b), *Northern Australia Insurance Premiums Taskforce: Final Report*, Commonwealth of Australia, *http://treasury.gov.au/~/media/Treasury/Consultations%20and%20Reviews/Reviews*

*%20and%20Inquiries/2015/NAIP%20Taskforce/Final%20Report/Downloads/PDF/NAIP_final_report.ashx.*

The Australian Government the Treasury (2011), *Natural Disaster Insurance Review: Inquiry into flood insurance and related matters (September 2011)*, Commonwealth of Australia, *www.ndir.gov.au/content/report/downloads/NDIR_final.pdf.*

Ball, T., A. Werritty and A. Geddes (2013), "Insurance and sustainability in flood-risk management: the UK in a transitional state", *Area*, Vol. 45 (3), pp. 266-272.

Barker, G. and R. Tooth (2008), "An Analysis of the Demand for House and Contents Insurance in Australia", *Centre for Law and Economics, http://law.anu.edu.au/cle/Papers/Insurance_Demand_Home&Contents_Fina.pdf.*

Bin, O. and C. Landry (2013), "Changes in implicit flood risk premiums: Empirical evidence from the housing market", *Journal of Environmental Economics and Management*, Vol. 65, pp. 361–376.

Botzen, W. J. W. et al. (2009), "Willingness of homeowners to mitigate climate risk through insurance", *Ecological Economics*. Vol. 68 (8–9), pp. 2265–2277.

Botzen, W.J.W., J. de Boer and T. Terpstra (2013), "Framing of risk and preferences for annual and multi-year flood insurance", *Journal of Economic Psychology*, Vol. 39, pp. 357–375.

Botzen, W.J.W., H. Kunreuther and E Michel-Kerjan (2015), "Divergence between individual perceptions and objective indicators of tail risks: Evidence from floodplain residents in New York City", *Judgement and Decision Making*, Vol. 10 (4), pp. 365-385.

Botzen, W.J.W., and J. C. J. M. van den Bergh (2009), "Managing natural disaster risks in a changing climate", *Environmental Hazards*, Vol. 8(3), pp. 209-225.

Boyle, C.E. (2015), "What the Future Holds for Catastrophe Modeling: Aon's Impact Forecasting", *Carrier Management*, 23 June.

Bukvic, A., A. Smith and A. Zhang (2015), "Evaluating drivers of coastal relocation in Hurricane Sandy affected communities", *International Journal of Disaster Risk Reduction*, Vol. 13, pp. 215–228.

CBO (2007), *Value of Properties in the National Flood Insurance Program*, The Congress of the United States, Congressional Budget Office, Washington, DC.

CII New Generation Programme Claims Group 2014 (2015), *Co-ordination of Flood Response*, The Chartered Insurance Institute, London, 18 May.

Collins, E. and L. Simpson (2007), "The impact of climate change on insuring flood risks", Presented to the Institute of Actuaries of Australia Biennial Convention, 23-26 September, Christchurch, New Zealand, *http://actuaries.asn.au/Library/1.d_Conv07_Paper_Collins%20Simpson_The%20impact%20of%20climate%20change%20on%20insuring%20flood%20risk.pdf.*

Comité Européen des Assurances (2005), *Flood Prevention in Europe: The Role of the Insurance Industry*, Comité Européen des Assurances, Paris.

Costanza, R. et al. (2008), "The value of coastal wetlands for hurricane protection", *Ambio*, Vol. 37, pp. 241–248.

Crichton, D. (2008), "Role of Insurance in Reducing Flood Risk", *The Geneva Papers on Risk and Insurance*, Vol. 33, The Geneva Association (The International Association for the Study of Insurance Economics).

DEFRA (2013), *Securing the future availability and affordability of home insurance in areas of flood risk*, Department for Environment, Food and Rural Affairs, London.

DEFRA (2012), *Obtaining flood insurance in high risk areas*, Department for Environment, Food and Rural Affairs, London.

Department of Communities and Local Government (2012), *National Planning Policy Framework*, Department of Communities and Local Government, London, *www.gov.uk/government/uploads/system/uploads/attachment_data/file/6077/2116950.pdf.*

Department of Finance Canada (2014), *Budget 2014: The Road to Balance: Creating Jobs and Opportunities*, Her Majesty the Queen in Right of Canada, *www.budget.gc.ca/2014/docs/plan/pdf/budget2014-eng.pdf.*

Douglas, J., M. Bowditch and A. Ni (2013), *Affordability of Natural Disaster Insurance*, Australian Centre for Financial Studies, *www.australiancentre.com.au/sites/default/files/NewsDocs/D1P4%20-%20Affordablity%20of%20natural%20disaster%20insurance%20Justin%20Douglas%20Treasury.pdf,* accessed 6 October 2015.

Ebert, S., O. Hulea and Strobel (2009), "Floodplain restoration along the lower Danube: A climate change adaptation case study", *Climate and Development*, Vol. 1, pp. 212-219.

Environment Agency (2014a), *Flood and Coastal Erosion Risk Management Outcome Measures: Q2 2014/15 (1st July - 30th September 2014)*, Environment Agency, London, *www.gov.uk/government/uploads/system/uploads/attachment_data/file/389952/FCERM_outcome__measures_Q2_2014_15__External.pdf,* accessed 12 January 2016.

Environment Agency (2014b), *Flood and Coastal Erosion Risk: Long-Term Investment Scenarios 2014*, Environment Agency, London, *https://www.gov.uk/government/uploads/system/uploads/attachment_data/file/381939/FCRM_Long_term_investment_scenarios.pdf,* accessed 6 June 2016.

Ernst & Young (2014), *2014 Global Consumer Insurance Survey: Reimagining customer relationships*, Ernst & Young, London, *www.ey.com/GL/en/Industries/Financial-Services/Insurance/ey-2014-global-customer-insurance-survey,* accessed 11 January 2016.

European Union (2007), Council Directive 2007/60/EC of the European Parliament and of the Council of 23 October 2007 on the assessment and management of flood risks, *Official Journal of the European Union*, 6.11.2007, Brussels.

Federal Emergency Management Agency (2015), *Fact sheet: Community Rating System*, Federal Insurance and Mitigation Administration, Department of Homeland Security (October), *www.fema.gov/media-library-data/1444399187441-5293d81167caaf062c2925b75a69215f/NFIP_CRS_Fact_Sheet-Oct-8-2015.pdf,* accessed 1 December 2015.

Federal Office of the Environment (2015a), "Spatial planning: Good Maps for Risk Management", *Living with Natural Hazards - Magazine «environment»*, Vol. 2, *www.bafu.admin.ch/naturgefahren/14144/15299/15324/index.html?lang=en,* accessed 29 March 2016.

Federal Office of the Environment (2015b), "10 years after 2005: We have made major gains in terms of safety", *Living with Natural Hazards - Magazine «environment»*, Vol. 2, *www.bafu.admin.ch/naturgefahren/14144/15299/15318/index.html?lang=en,* accessed 29 March 2016.

Fink, A., U. Ulbrich and H. Engel (1996), "Aspects of the January 1995 flood in Germany", *Weather*, Vol. 51, pp. 34–39.

Flood Re (2015), How it works (website), *www.floodre.co.uk/how-it-works-overview,* accessed 12 November 2015.

Gallagher, J. (2014), "Learning about an Infrequent Event: Evidence from Flood Insurance Take-Up in the United States", *American Economic Journal: Applied Economics*, Vol. 6 (3), pp. 206-233.

Galloway, G. E. (1995), "Learning from the Mississippi Flood of 1993: Impacts, Management Issues, and Areas for Research", *U.S.-Italy Research Workshop on the Hydrometerology, Management, and Impacts of Extreme Floods*, Perugia, Italy.

GAO (2014), *Flood Insurance: Strategies for Increasing Private Sector Involvement (GAO-14-127)*, Government Accountability Office (January), Washington, DC.

GAO (2008), *Flood Insurance: FEMA's Rate-Setting Process Warrants Attention (GAO-09-12)*, Government Accountability Office, Washington, DC.

GDV (2013), *Position Paper of the German Insurance Association (GDV) on the consultation on the Green Paper on the Insurance of Natural and Man-Made Disasters*, German Insurance Association (Gesamtverband der Deutschen Versicherungswirtschaft), Berlin.

General Accounting Office (2004), "National Flood Insurance Program: Actions to Address Repetitive Loss Properties - Statement of William O. Jenkins, Jr., Director Homeland Security & Justice Issues", *Testimony Before the Subcommittee on Economic Policy, Committee on Banking, Housing, and Urban Affairs, U.S. Senate (GAO-04-401T)*, General Accounting Office, Washington, DC, 25 March.

Goodnough, A. (2006), "As Hurricane Season Looms, States Aim to Scare", *The New York Times*, May 31.

Green, C. and E. Penning-Roswell (2004), "Flood Insurance and Government: "Parasitic" and "Symbiotic" Relations", *The Geneva Papers on Risk and Insurance*, Vol. 29 (3), The Geneva Association (The International Association for the Study of Insurance Economics).

Grothmann, T. and F. Reusswig (2006), "People at Risk of Flooding: Why Some Residents Take Precautionary Action While Others do not", *Natural Hazards*, Vol. 38, pp. 101–120.

Hartwig, R. and C. Wilkinson (2014), *Residual Market Property Plans: From Markets of Last Resort to Markets of First Choice*, Insurance Information Institute, New York.

Hallegatte, S. (2012), "A Cost-Effective Solution to Reduce Disaster Losses in Developing Countries: Hydro-Meteorological Services, Early Warning, and Evacuation", *Policy Research Working Paper 6058*, World Bank, Washington, DC.

Healy, A., and N. Malhotra (2009), "Citizen Competence and Government Accountability: Voter Responses to Natural Disaster Relief and Preparedness Spending", *http://myweb.lmu.edu/ahealy/papers/healy_prevention_070808.pdf,* accessed 20 November 2015.

Henrich, L., J. McClure and M. Crozier (2015), "Effects of risk framing on earthquake risk perception: Life-time frequencies enhance recognition of the risk", *International Journal of Disaster Risk Reduction*, Vol. 13, pp. 145-150.

Huber, M. (2012), *Fixing a Broken National Flood Insurance Program: Risks and Potential Reforms*, Center for Climate and Energy Solutions.

Huber, M. (2004), "Insurability and Regulatory Reform: Is the English Flood Insurance Regime able to Adapt to Climate Change?", *The Geneva Papers on Risk and Insurance*, Vol. 29 (2), The Geneva Association (The International Association for the Study of Insurance Economics).

Hudson, P. et al. (2014), "Risk Selection and Moral Hazard in Natural Disaster Insurance Markets: Empirical evidence from Germany and the United States", *Wharton Risk Center Working Paper #2014–07.*

IFRC (2001), *World Disasters Report*, International Federation of Red Cross and Red Crescent Societies, Geneva.

Insurance Bureau of Canada (2015), *The financial management of flood risk: An international review – lessons learned from flood management programs in G8 countries*, IBC, Toronto.

International Commission for the Protection of the Rhine (2002), *Non structural flood plain management – measures and their effectiveness.*

Jarraud, M. (2007), *WMO Secretary-General Opening Statement of Madrid Conference: Social and Economic Benefits of Weather-, Climate- and Water-Related Information and Services*, Madrid, 19-22 March.

Johnson, E. and D. Goldstein (2003), "Do Defaults Save Lives?" *Science*, Vol. 302, pp. 1338-1339.

Jongejan, R. and P. Barrieu (2008) "Insuring Large-Scale Floods in the Netherlands", *The Geneva Papers on Risk and Insurance Issues and Practice*, Vol. 33, The Geneva Association (The International Association for the Study of Insurance Economics).

Kaplan, T. (2013), "Cuomo seeking home buyouts in flood zones", *The New York Times*, 3 February.

Kay, R., and I. Wilderspin (2002). "Box 4.4: Mangrove Planting Saves Lives and Money in Vietnam." in *World Disaster Report Focus on Reducing Risk*, International Federation of Red Cross and Red Crescent Societies, Geneva.

Keating, A. et al. (2014), *Operationalizing Resilience against Natural Disaster Risk: Opportunities, Barriers, and a Way Forward*, Zurich Flood Resilience Alliance (Zurich Insurance Company Ltd.), Zurich, Switzerland.

Kidney, S. (2016), "Green Bond Market Report: Four new 'AAA' rated benchmark sized green bonds: Kommuninvest inaugural $600m (3yr) NWB Bank $1bn (10yr) Chinese bank SPD RMB 15bn ($2.3bn, 5yr) IFC $700m (10yr); &1st ever green Schuldschein frm Nordex €550m is CB Cert", *Climate Bonds Initiative* (website), *www.climatebonds.net/2016/04/green-bond-market-report-four-new-%E2%80%98aaa%E2%80%99-rated-benchmark-sized-green-bonds-kommuninvest,* accessed at 11 April 2016.

Knowles, G. and H. Kunreuther (2014), "Troubled Waters: The National Flood Insurance Program in Historical Perspective", *Journal of Policy History*, Vol. 26 (3), pp. 327-353.

Kousky, C. and H. Kunreuther (2014), "Addressing Affordability in the National Flood Insurance Program", *Journal of Extreme Events*, Vol. 1 (1).

Kousky, C. and E. Michel-Kerjan (2015), "Examining flood insurance claims in the United States: six key findings", *The Journal of Risk and Insurance (forthcoming).*

Kousky, C. and L. Shabman (2015), *A Proposed Design for Community Flood Insurance*, Resources for the Future, Washington, DC, December.

Kreibich, H. et al. (2005), "Flood loss reduction of private households due to building precautionary measures: Lessons learned from the Elbe flood in August 2002", *Natural Hazards and Earth System Sciences*, Vol. 5(1), pp. 117-126.

Kunreuther, H. et al. (2008), *Managing Large-Scale Risks in a New Era of Catastrophes: Insuring, Mitigating and Financing Recovery from Natural Disasters in the United States*, Wharton Risk Management and Decision Processes Center/Georgia State University and Insurance Information Institute, Philadelphia.

Kunreuther, H. and E. Michel-Kerjan (2013a), "Implementing the National Flood Insurance Reform Act in a New Era of Catastrophes", *Penn Wharton Public Policy Initiative Issue Brief*, Vol. 1 (9).

Kunreuther, H. and E. Michel-Kerjan (2013b), "Policy Options for Reducing Losses from Natural Disasters: Allocating $75 billion", *Natural Disasters: Challenge Paper*, Copenhagen Consensus 2012.

Kunreuther, H. and E. Weber (2014), "Aiding Decision Making to Reduce the Impacts of Climate Change", *Journal of Consumer Policy*, Vol. 37, pp. 397-411.

Landry, C. and M. Jahan-Parvar (2011), "Flood Insurance Coverage in the Coastal Zone", *The Journal of Risk and Insurance*, Vol. 78 (2), pp. 361-388.

Lyles, W., P. Berke and G. Smith (2014), "Impacts of Federal and State Hazard Mitigation Policies on Local Land Use Policy", *Journal of Planning, Education and Research*, Vol. 34(1), pp. 60-76.

MacClune et al. (2015), *Urgent case for recovery: what we can learn from the August 2014 Karnali River floods in Nepal*, Zurich Flood Resilience Alliance (Zurich Insurance Company Ltd.), Zurich, Switzerland.

McIvor, A. L. et al. (2012), "Reduction of Wind and Swell Waves by Mangroves." *Natural Coastal Protection Series: Report 1/Cambridge Coastal Research Unit Working Paper 40*, The Nature Conservancy and Wetlands International.

Meyer, E. et al. (2013), "The dynamics of hurricane risk perception: Real-Time Evidence from the 2012 Atlantic Hurricane Season", *Wharton Risk Center Working Paper #2013–07*.

Michel-Kerjan, E., S. Lemoyne de Forges and H. Kunreuther (2012), "Policy Tenure under the U.S. National Flood Insurance Program (NFIP)", *Risk Analysis*, Vol. 32 (4).

Michel-Kerjan, E. (2010), "Catastrophe Economics: The National Flood Insurance Program", *Journal of Economic Perspectives*, Vol. 24 (4), pp. 165-186.

Miranda, J. (2013), "A Reinsurer's Perspective on the Current Catastrophe Modeling Landscape", *Carrier Management*, 30 September.

Mitchell, M. (2015), "Relocation after disaster: Engaging with insured residential property owners in Greater Christchurch's land-damaged 'Residential Red-Zone'", *Brookings-LSE Project on Internal Displacement*, Brookings Institution, Washington, DC.

MMC (2005), *Natural hazard mitigation saves: an independent study to assess the future savings from mitigation activities: Volume 1 – Findings, Conclusions, and Recommendations*, Multi hazard Mitigation Council, National Institute of Building Sciences, Washington, DC.

Munich Re (2012), *50th anniversary of storm surge in Hamburg: Subsequent flood controls prevent billions in losses (Press Release)*, Munich Reinsurance Company, Munich, *www.munichre.com/en/media-relations/publications/press-releases/2012/2012-02-13-press-release/index.html*, accessed 25 March 2016.

Nance, E. (2015), "Exploring the impacts of flood insurance reform on vulnerable communities", *International Journal of Disaster Risk Reduction*, Vol. 13, pp. 20-36.

National Research Council (2015), *Affordability of National Flood Insurance Program Premiums: Report 1*, The National Academies Press, Washington, DC.

National Academies of Sciences, Engineering, and Medicine (2015), *Affordability of National Flood Insurance Program Premiums: Report 2*, The National Academies Press, Washington, DC.

OECD (2016), *Mitigating Droughts and Floods in Agriculture: Policy Lessons and Approaches*, OECD Studies on Water, OECD Publishing, Paris, *http://dx.doi.org/10.1787/9789264246744-en.*

OECD (2015a), *Disaster Risk Financing: A global survey of practices and challenges*, OECD Publishing, Paris, *http://dx.doi.org/10.1787/9789264234246-en.*

OECD (2015b), *OECD Business and Finance Outlook 2015*, OECD Publishing, Paris. *http://dx.doi.org/10.1787/9789264234291-en.*

OECD (2014a), *Seine Basin, Île-de-France, 2014: Resilience to Major Floods*, OECD Publishing, Paris, *http://dx.doi.org/10.1787/9789264208728-en.*

OECD (2014b), *Water Governance in the Netherlands: Fit for the Future?*, OECD Studies on Water, OECD Publishing, Paris, *http://dx.doi.org/10.1787/9789264102637-en.*

OECD (2013a), *Water and Climate Change Adaptation: Policies to Navigate Uncharted Waters*, OECD Studies on Water, OECD Publishing, *http://dx.doi.org/10.1787/9789264200449-en.*

OECD (2013b), "Design and Delivery of Defined Contribution (DC) Pension Schemes: Policy challenges and recommendations", Presented at the Conference on Defined Contribution, Guarantees and Risk Sharing, 5 March 2013, London, United Kingdom, *www.oecd.org/daf/fin/private-pensions/DCPensionDesignHighlights.pdf,* accessed 12 April 2016.

OECD (2012), *Disaster Risk Assessment and Risk Financing: A G20/OECD Methodological Framework,* *www.oecd.org/daf/fin/insurance/g20oecdframeworkfordisasterriskmanagement.htm.*

Orie, M. and W. Stahel (eds.) (2013), "Insurers' contributions to disaster reduction—a series of case studies", *The Geneva Reports Risk and Insurance Research No. 7*, The Geneva Association (The International Association for the Study of Insurance Economics).

Owens, C. (2014), "A New Breed of Muni Bond Is Financing Climate Change Adaptation", *Next City: Inspiring better cities*, 26 September, *https://nextcity.org/daily/entry/green-bonds-municipal-bonds-resilience,* accessed 10 December 2015.

Palm, R. et al. (1990), *Earthquake Insurance in California: Environmental Policy and Individual Decision Making*, Westview Press, Boulder, CO.

Pasterick, E. (1998), "The National Flood Insurance Program", in *Paying the Price: The Status and Role of Insurance Against Natural Disasters in the United States*, H. Kunreuther and R. Roth (eds.), The National Academies Press, Washington, DC.

Pitt, Michael (2008), *Pitt Review: Learning lessons from the 2007 floods*, Cabinet Office, London.

Productivity Commission (2014), *Natural Disaster Funding Arrangements, Inquiry Report no. 74*, Canberra.

Property Casualty Insurers Association of America (2011), *True market risk rates for flood insurance*, Property Casualty Insurers Association of America, Des Plaines, Illinois, June.

Quantum Market Research (2013), *The Understand Insurance Research Report*, *http://understandinsurance.com.au/assets/pdf/FINAL%20Understand%20Insurance%20Research%20Report.pdf,* accessed 11 January 2016.

Ranger, N. et al. (2011), "An assessment of the potential impact of climate change risk in Mumbai", *Climatic Change*, Vol. 104, pp. 139–167.

Risk Management Solutions and Lloyd's (2008), *Coastal Communities and Climate Change: Maintaining Future Insurability*, Lloyd's, London.

Sandink, D. et al. (2010), *Making Flood Insurable for Canadian Homeowners: A Discussion Paper*, Institute for Catastrophic Loss Reduction & Swiss Reinsurance Company Ltd., Toronto.

Sarmiento, C. and T. Miller (2006), *Costs and Consequences of Flooding and the Impact of the National Flood Insurance Program: Prepared as part of the 2001-2006 Evaluation of the National Flood Insurance Program*, Pacific Institute for Research and Evaluation, Carleton, MD.

Schwarze et al. (2011), "Natural Hazard Insurance in Europe: Tailored Responses to Climate Change are Needed", *Environmental Policy and Governance*, Vol. 21, pp. 14-30.

Schwarze, R. and G. Wagner (2007), "The Political Economy of Natural Disaster Insurance: Lessons from the Failure of a Proposed Compulsory Insurance Scheme in Germany", *European Environment*, Vol. 17, pp. 403–415.

Siders, A. (2013), "Hurdles for Gov. Cuomo's New York buyout program", *Climate Law Blog*, Sabin Center for Climate Change Law, *http://blogs.law.columbia.edu/climatechange/2013/02/05/hurdles-for-gov-cuomos-new-york-buyout-program/*, accessed 20 November 2015.

Simpson, A. (2014), "With Premium Refunds in Process, Criticism of Flood Insurance Program Ebbs", *Carrier Management*, 28 July.

Stelmakowich, A. (2013), "Overland Overhaul", *Canadian Underwriter*, August, *www.canadianunderwriter.ca/news/overland-overhaul/1002551174/?type=Print%20Archives*, accessed 11 January 2016.

Suncorp Insurance (2013), *Suncorp starts writing new business as Roma levee kicks off: Estimated premium reductions of up to 80% upon completion (Press release)*, Suncorp Insurance Company, 24 September, *www.suncorpgroup.com.au/sites/default/files/pdf/news/Media%20Release%20and%20Infographic%20-%20Roma%20sod%20turning%20-%20FINAL.pdf*, accessed 29 March 2016.

Surminski et al. (2014), "Reflection on the current debate on how to link flood insurance and disaster risk reduction in the European Union", *Centre for Climate Change Economics and Policy Working Paper No. 184/ Grantham Research Institute on Climate Change and the Environment Working Paper No. 162*, Centre for Climate Change Economics and Policy, Leeds.

Surminski, S. and J. Eldridge (2014), "Flood insurance in England – an assessment of the current and newly proposed insurance scheme in the context of rising flood risk", *Centre for Climate Change Economics and Policy Working Paper No. 161/Grantham Research Institute on Climate Change and the Environment Working Paper No. 144*, Centre for Climate Change Economics and Policy, Leeds.

Swiss Re (2015), *The natural catastrophe protection gap in Italy: time for action*, Swiss Reinsurance Company Ltd., Zurich.

Swiss Re (2012), *Flood – an underestimated risk: Inspect, Inform, Insure*, Swiss Reinsurance Company Ltd., Zurich.

Swiss Re (2010), *Weathering climate change: Insurance solutions for more resilient communities*, Swiss Reinsurance Company Ltd., Zurich.

Thieken, A. et al. (2006), "Insurability and Mitigation of Flood Losses in Private Households in Germany", *Risk Analysis*, Vol. 26 (2).

Thistlethwaite, J. and B. Feltmate (2013), *Assessing the Viability of Overland Flood Insurance: The Canadian Residential Property Market*, The Co-operators.

Troy, A. and J. Romm (2004), "Assessing the Price Effects of Flood Hazard Disclosure under the California Natural Hazard Disclosure Law (AB 1195)", *Journal of Environmental Planning and Management*, Vol. 47 (1), pp. 137–162.

US Green City Bonds Coalition (2015), *How to issue a Green Muni Bond: The Green Muni Bonds Playbook*, Climate Bonds Initiative, *www.climatebonds.net/files/files/Green%20City%20Playbook.pdf,* accessed 11 April 2016.

Vajjhala, S. and J. Rhodes (2015), *Leveraging catastrophe bonds as a mechanism for resilient infrastructure project finance*, Rebound: Insuring for Resilience, *www.refocuspartners.com/reports/RE.bound-Program-Report-December-2015.pdf,* accessed 10 December 2015.

Vellinga, P. et al. (2009), "Exploring High-End Climate Change Scenarios for Flood Protection of the Netherlands", *KNMI Scientific Report 2009-05*, De Bilt, Netherlands.

Ward, S. (2015), "Importance of a holistic approach to financial management of disaster risks-Australia's experience", Presented to the Global Seminar on Disaster Risk Financing: Towards the development of effective approaches to the financial management of disaster risks, 17-18 September, Kuala Lumpur, Malaysia, *www.oecd.org/daf/fin/insurance/global-seminar-disaster-risk-financing-2015.htm.*

Warner, K. et al (2009), *Adaptation to Climate Change: Linking Disaster Risk Reduction and Insurance*, United Nations International Strategy for Disaster Reduction Secretariat, Geneva, *www.unisdr.org/files/9654_linkingdrrinsurance.pdf.*

Wassell, P. et al. (2009), "Resilient Reinstatement: The Costs of flood resilient reinstatement of domestic properties", *ABI Research Paper No 14*, Association of British Insurers. London.

Wilby, R. and R. Keenan (2012), "Adapting to flood risk under climate change", *Progress in Physical Geography*, Vol. 36 (3), pp. 348–378.

World Bank and Thai Ministry of Finance (2012), *Rapid Assessment for Resilient Recovery and Reconstruction Planning*, World Bank, Bangkok.

Zurich Insurance Company (2014), *Central European floods 2013: a retrospective*, Zurich Flood Resilience Alliance (Zurich Insurance Company Ltd.), Zurich, Switzerland.

*Chapter 5*

# Managing the fiscal cost of floods

*This chapter provides an overview of the potential fiscal costs of floods and approaches to managing those costs, including ways to minimise the costs to the public sector for providing compensation and public insurance as well as mechanisms for transferring public sector risk to insurance and capital markets.*

In flood-prone countries, governments (local, regional and national) face significant costs related to the financial management of flood risk. This includes both the costs of investing in *ex ante* risk reduction as well as *ex post* costs related to emergency response, reconstruction of public assets, and compensation and financial assistance to sub-national governments, businesses and individuals affected by floods. Where governments provide insurance, reinsurance or guarantees, they may also face costs where covered losses exceed any accumulated reserves to manage those losses. In extreme cases, governments may also face unanticipated costs as a result of gaps in the financial (risk-bearing) capacity of households, businesses, banks, insurance companies and/or other levels of government to manage the losses they would normally be expected to absorb (OECD, 2012).

Effective financial management of these fiscal costs requires governments to: i) assess the potential exposures that they face, based on a range of potential flood scenarios, both normal and extreme; and ii) evaluate the most effective way to manage these exposures, considering the potential roles of investments in risk reduction, risk transfer and *ex post* response and taking into account the potential deterioration in macro-economic conditions that may occur in the event of severe flooding.

## 5.1 The fiscal costs of floods

The financial management of flood risk can create significant costs for governments in terms of investments in risk reduction as well as costs related to emergency response and reconstruction. In the United States, for example, federal government payments for natural disaster assistance (for all perils, not just floods) were approximately USD 30-40 billion per year between 2005 and 2013 (Conrad and Thomas, 2013; Weiss and Weidman, 2013) – more than the USD 30 billion that has been paid out annually in insured losses. There is some evidence that federal spending is accounting for an increasing share of losses (23% for Hurricane Hugo in 1989, 50% for Hurricane Katrina in 2005 and over 80% for Hurricane Sandy in 2012) (Kunreuther and Kerjan, 2013). In Canada, flood assistance has accounted for 70-80% of all federal assistance provided under the Disaster Financial Assistance Arrangements and has cost approximately CAD 3.7 billion per year between 2010 and 2013 (Insurance Bureau of Canada, 2015). In Italy, government expenditures on emergency response and reconstruction related to hydrological events has been estimated at EUR 2.6 billion per year between 2010 and 2012 (Swiss Re, 2015). In Central Europe (Slovak Republic, Poland, Czech Republic and Hungary), a 1-in-20 year (5% annual probability) flood could lead to losses equivalent to 2.2-10.7% of government revenues (Pollner, 2012). For some countries, particularly developing countries, the impact of a large flood event could have a significant impact on public finances and even on sovereign credit ratings as a result of a reduction in economic growth (affecting revenue), increases in public spending on reconstruction and a deterioration in export performance. As noted in Box 2.2, a 1-in-250 year flood event could lead to a sovereign credit rating downgrade of 1.62 notches (and 3.41 notches in the context of climate change) in Thailand (Standard and Poor's Ratings Services, 2015).

In general, the most significant costs relate to *ex post* response, recovery and reconstruction, including emergency response costs, reconstruction of public assets and compensation and financial assistance to sub-national governments, businesses and individuals. National governments often incur a significant share of the costs of restoration and reconstruction of public assets owned by sub-national levels of government through compensation arrangements (even though, in many countries, ta significant share of public assets are owned by regional and/or local government (for example, 90% in the case of Australia (Productivity Commission, 2014)). For example, the largest share of the USD 60 billion in disaster relief funds appropriated in response to Hurricane Sandy in the United States went to programmes supporting the rehabilitation of public buildings and infrastructure (compared to USD 7.8 billion in claims payments from the NFIP for damage to public property (National Academy of Sciences, 2015b)).

In Australia, Canada and New Zealand, compensation is provided to sub-national governments based on a cost-sharing formula that varies based on the amount of sub-national government expenditure relative to financial capacity (with the federal share rising as the burden on sub-national governments increases). These programmes reimburse a share of eligible expenses incurred by sub-national governments for costs such as emergency response, restoration and reconstruction of public and (sometimes private) assets. Austria also provides financial assistance to sub-national levels of government under the *Catastrophe Fund Act*, including to share the costs of restoring public and private assets. In the United States, the federal government is legally obligated to offer assistance when state and local resources are overwhelmed as a result of a catastrophe.

Financial assistance may also be provided to businesses and individuals for loss of income or more generally to support economic recovery. In Australia, for example, the national government can provide an Australian Government Disaster Recovery Allowance (to address lost income) and an Australian Government Disaster Recovery Payment (to support recovery) (OECD, 2015). In countries with public insurance schemes, the *ex post* cost of a flood event will depend on the structure of the programme, and particularly, the extent to which the premiums charged reflect the true actuarial cost of the coverage. To calculate the potential fiscal costs of flood events, countries need to estimate the probability and impact of an event with financial impacts exceeding the capacity of the public insurance scheme to absorb. In countries that provide a backstop on losses to private insurance (e.g. Belgium), the *ex post* cost could be calculated as the government's potential payments to insurers as a function of the probability of an event expected to trigger the guarantee (up to any established limit).

## 5.2 Minimising fiscal costs

There are a number of approaches to managing the fiscal costs of floods. For example, cost-efficiency gains may be possible in terms of emergency preparedness or response. As outlined in Chapter 4, overall losses can be reduced through investments in risk reduction at the community and household level. In addition, the design and scope of financial assistance and compensation arrangements (and public insurance schemes, where applicable) can have important implications on the overall costs to the public sector. Finally, the relative share of costs that is expected to be covered by private businesses and individuals and the share covered by private insurance or capital markets will also have important fiscal implications.

### *Compensation and financial assistance schemes*

Post-disaster compensation and financial assistance can make an important contribution to reducing hardship and supporting recovery among those seriously affected, particularly where alternative forms of financial protection are unavailable or unaffordable for the most vulnerable segments of society. After large flood events, there will generally be a significant delay between the occurrence of the event and the payment of insurance claims. Indemnity-based insurance payments require loss adjustment which will usually be hampered by problems with accessibility to the affected area as well as the availability of sufficient adjustors in the face of large amounts of claims. As a result, timely financial assistance to support immediate basic needs, such as short-term accommodation, food and clothing, can be critical as a post-disaster response for reducing the impact on vulnerable households (Productivity Commission, 2014). Financial assistance can also be provided to reduce the impact of a temporary loss of income or support some of the costs of reconstruction of private property. However, the scope of assistance available will have important implications in terms of the incentives created for securing financial protection and implementing risk reduction measures which needs to be accounted for when considering the overall cost of such arrangements.

In many countries, the scope of potential *ex post* financial assistance is not defined *ex ante*. A lack of *ex ante* clarity on the level of financial assistance that may be available can create expectations and misunderstandings in terms of the amount of assistance that will be available to individual households (as outlined in section 4.3.1) as well as room for political discretion in terms of the amount of assistance eventually granted (where politicians will have an incentive to ensure their constituents are well-compensated –a likely factor in the increasing levels of government assistance seen in many countries over time).

Clear definition of the scope and eligibility for *ex post* financial assistance, applied consistently over time and across disaster events, combined with a narrow focus on providing for immediate needs, would reduce misunderstanding and the potential for adverse impacts on incentives for self-protection. In some countries (e.g. Netherlands), statutory ceilings on the maximum amount of government compensation have been established (although the effectiveness of these ceiling in limiting fiscal costs in the face of a significant event may be limited where the ceiling may be amended). Clear definitions on scope and eligibility for assistance will also provide a clearer understanding of the government's potential exposure to financial assistance and compensation payments. A lack of specificity in eligibility terms and payment amounts (and/or too much scope for political discretion) can lead to unexpected costs after an event. For example, the Australian Government Disaster Recovery Payment was initially estimated to cost AUD 3 million per year but payments reached AUD 850 million following the Queensland floods and Cyclone Yasi in 2010-2011 (Productivity Commission, 2014). In countries where different levels of government have the authority to provide financial assistance, it will also be important to ensure coordination and avoid duplication. There may be advantages in placing responsibility for financial assistance with the national government in order to ensure equivalent treatment for all citizens, irrespective of the local administrative area in which they reside (Productivity Commission, 2014).

To reduce the impact of financial assistance and compensation on insurance demand, a number of countries do not provide compensation for damages that would have otherwise been insurable. Importantly, such a limitation should be focused on what could have been insured rather than what was actually insured as compensation for uninsured damages resulting from a reluctance by the owner to seek insurance coverage could create moral hazard and reduce the demand for insurance. In the German Federal State of Saxony, for example, eligibility for payment of post-disaster financial assistance explicitly excludes cases where an affected person has failed to take the necessary steps to prevent a need for assistance, including acquisition of insurance (where it can be done at an economically acceptable price) (GDV, 2013). Higher levels of insurance coverage for flood risk, supported as necessary by measures to support insurability (as described in

Chapter 4), should reduce the "financial protection gap" and the resulting level of political pressure for government compensation. In countries where insurance coverage for flood is not available, compensation could be restricted in cases where homeowners had an opportunity to protect against future floods (e.g. through government funding for risk reduction measures) but chose not to - as proposed by the government of Alberta (Canada) after 2013 flooding in Calgary (Stelmakowich, 2013).

Consideration could also be given to the eligibility of businesses for receiving *ex post* disaster assistance. Many countries provide grants or loans to businesses affected by disasters with the aim of supporting business survival in the face of severe events (and therefore employment). For example, eligible expenses under Australia's National Disaster Relief and Recovery Arrangements (NDRRA) include loans and grants for small businesses affected by disaster events. Small businesses can face significant losses in income after disaster events as a result of disruptions to critical infrastructure services such as water or power and disruptions to important clients or suppliers. The US Insurance Institute for Business & Home Safety estimates that 25% of small businesses never reopen after a hurricane, flood, wildfire or other catastrophic event. However, there is limited evidence that such assistance has significant benefits in terms of improving business survival (Regional Australia Institute, 2014). It may be more effective to focus on measures to improve insurance penetration (including business interruption coverage) among small business rather than providing *ex post* compensation. A recent survey of small businesses in the United States, for example, found that approximately two-thirds of small businesses did not have business interruption insurance (Nationwide, 2015). Similarly, the recent flooding in the United Kingdom revealed that many small and family-run businesses did not have commercial insurance coverage at all (Cohn and Holton, 2015). In addition to providing financial protection, commercial insurance providers in most countries will generally charge risk-based premiums and provide risk management advice to their clients which should support risk reduction (and ultimately reduce business disruption).

Similar to the case of individuals, financial assistance and compensation to sub-national governments can be crucial for ensuring their ability to manage the financial impacts of significant events when their risk-bearing capacity is insufficient. The amount of national assistance provided to sub-national governments should take into account the relative fiscal capacity between national and sub-national levels of government (share of national revenues relative to the share of public spending responsibilities) (Productivity Commission, 2014). However, efforts should also be made to ensure that national compensation does not reduce the incentive for sub-national governments to invest in risk reduction and/or financial protection for sub-national assets – particularly given the important responsibilities that local governments usually have in terms of land-use planning and resilience of public infrastructure. In Australia and New Zealand, where national governments compensate sub-national governments for 75% and 60% of the cost of public infrastructure reconstruction respectively, there is some evidence that these arrangements have impeded the take-up of insurance by sub-national governments (Productivity Commission, 2014). In Australia, the Attorney General has the authority to review the insurance arrangements of state governments, make recommendations on changes to those arrangements and penalise states that do not comply with its recommendations, including through reductions to the rate of reimbursement for reconstruction costs. The Congressional Research Service has put forward a similar proposal in the United States, suggesting that the 75% reimbursement rate be reduced in states that don't adhere to land-use restrictions and building codes (Lindsay and

McCarthy, 2015). The significant involvement of national governments in local reconstruction may also reduce market-based incentives for risk reduction, such as credit ratings and borrowing costs. For example, in the United States, the extent of federal *ex post* assistance (which reimburses 80% to 100% of all expenses) has been identified as a possible reason why disaster risks are not considered in credit ratings for municipal bonds (Tournier, 2011).

One approach may be to establish compensation arrangements between national and regional/local governments as an insurance arrangement with premiums potentially varying by level of risk, rates of co-insurance and coverage limits (Productivity Commission, 2014). A given state or local government could determine the level of coverage that they seek and the share of the reconstruction costs that they are willing to bear and then pay a premium to the national government for providing that coverage. The national government could encourage risk reduction at the level of state or local government by varying the premium charged with risk, providing discounted premiums in response to risk reduction measures and/or ensuring that there is a material level of co-insurance/cost-sharing embedded in the compensation arrangement. Depending on the specific arrangements, there is some similarity between this approach and the approach taken by countries with a government insurer of local public assets (such as the Philippines or Indonesia – see section 5.3 below).

As noted above, a variation on this approach is being examined in the United States in the form of community-based flood insurance. The *Homeowner Flood Insurance Affordability Act of 2014* asked FEMA to evaluate the prospects for a community-based approach to flood insurance, culminating in a report by a Committee on Community-Based Flood Insurance Options within the National Academy of Sciences (National Academy of Sciences, 2015b). Community-based flood insurance would arrange insurance at the community-level (rather than at the level of individual households) with coverage that includes all private and public assets in the community (or all public and private properties with owners that wish to participate). Communities would be responsible for collecting funds from residents for the payment of premiums (to a government insurer, such as the NFIP, or a private insurance company) which could be risk-based, thereby encouraging communities to invest in risk reduction in order to benefit from lower premiums.

Focusing national compensation on "betterment" (i.e. improving the resilience of structures against flood risk) could also provide positive incentives for flood risk reduction. Financial assistance for betterment needs to be designed to ensure that there are no significant disadvantages to rebuilding at a higher level of protection. For example, in Australia, the Productivity Commission (2014) identified a number of barriers to accessing national funding for betterment, including a lower cost-sharing rate for betterment investments relative to the amount available for construction to pre-existing standards.

### *Public insurance schemes*

Where governments provide direct insurance, reinsurance or guarantees for flood (or disaster) risks, fiscal costs can be managed by ensuring that the insurance facility is self-sustaining and by maximising the share of risks covered by private insurance markets.

To be self-sustaining, a public insurance scheme must collect sufficient revenue in terms of premium income and fees to cover claims and administrative expenses. This requires premiums to be risk-based (at least in aggregate) and able to cover administrative costs. As noted in Table 3.1, a number of countries (including Canada, Costa Rica, France, Hungary, Iceland, Israel, Latvia, New Zealand, Peru, Philippines, Spain, Switzerland, United Kingdom and the United States) offer some form of public flood insurance or reinsurance for motor vehicles, residential properties/buildings and/or commercial properties/buildings. In some cases (e.g. Canada (motor vehicles), Costa Rica), the public insurance is provided for multiple perils (including non-catastrophe perils) through a public insurance company. In other cases, the public insurance or reinsurance was established specifically to provide coverage for catastrophe-related risk (France, Iceland, New Zealand, Spain, Switzerland) or flood risk in particular (Hungary, United Kingdom, United States).

Most of the public insurers providing coverage for flood risk have been designed to be generally self-sustaining. For example, in Spain, the CCS has operated for over 60 years without requiring funding from the government for losses beyond its capacity to pay. In Iceland, between 2004 and 2014, ICI collected EUR 117 million and faced claims costs of EUR 74 million for the coverage of all catastrophe risks (of which flooding accounted for approximately 4% of claims). In France, the government guarantee for losses of the CCR has been triggered infrequently, including after large losses due to flooding and windstorms in 1999 (Michel-Kerjan, 2001). In the United Kingdom, Flood Re has been designed to be self-sustaining up to a specific level of risk (i.e. a 1-in-200 year flood event with expected losses of GBP 2.1 billion). Premiums for Flood Re policies (and surcharges for other policyholders to cross-subsidise Flood Re policies) have been established specifically to provide sufficient reserves (and/or funding for reinsurance premiums) to manage that level of losses (DEFRA, 2013).

However, self-sustainability will not always be consistent with the goal of providing affordable insurance coverage, particularly in the context of an extraordinary event. For a public insurer subject to political pressure, building sufficient reserves to absorb very low-frequency, high-impact events may not be feasible given that a large unused reserve will be perceived as a sign that premiums are too high. In the United States, NFIP premiums are established based on average annual losses exclusive of extraordinary events and therefore the NFIP has borrowed funds from the US Treasury to address extraordinary losses from Hurricane Katrina in 2005 and Hurricane Sandy in 2012 (see Box 5.1). This is consistent with the original intention for the US Treasury to act as a "reinsurer" (with no premium collected) for extraordinary NFIP losses (National Research Council, 2015) (the *Biggert–Waters Flood Insurance Reform Act of 2012* has authorised the NFIP to seek reinsurance from the private market).

One means of minimising the fiscal cost of a public insurance scheme (and supporting affordability) is to ensure that the scheme is administered as efficiently as possible. In the United States, nearly all NFIP policies are distributed by participating private insurance companies, under the Write-Your-Own (WYO) program, where insurers are responsible for writing and servicing standard NFIP policies. Under this arrangement, the insurance companies benefit in terms of being able to offer their clients a wider range of coverage while the NFIP benefits from access to the insurance companies' marketing and distribution channels and claims settlement expertise. This program has been criticised for being overly generous to the distributing insurers (an estimated 30% of collected premiums is paid to Write-Your-Own insurers for claims adjustment and processing expenses (Michel-Kerjan, 2010)) although this is generally consistent with the ceding commission paid to direct insurers by reinsurers to cover administrative costs. An examination by the US GAO (2010) of the operations of six companies found that payments to the insurers were about 16.5% more than actual expenses incurred.

## Box 5.1. NFIP funding deficit

The establishment of NFIP premiums based on average annual losses exclusive of extraordinary events as well as the premium subsidies provided to pre-FIRM, "grandfathered" and other policyholders (estimated to cost USD 1.3 billion annually in premiums not collected for properties that benefit from premium rates below the true actuarially-sound rate (CBO, 2006)) has resulted in a funding deficit over time that has been managed through borrowing from the US Treasury. As can be seen in Figure 5.1 below, while the NFIP has attained a premium surplus (i.e. premiums collected were greater than claims payments made) in most years, with the exception of 1995, 2005 (and 2012 – not shown), the surplus of premiums collected has not allowed for the accumulation of sufficient reserves to address extreme loss years (consistent with the NFIP's approach to pricing). In extreme loss years, the NFIP has borrowed funds from the US Treasury, including USD 18 billion after Hurricane Katrina and approximately USD 7 billion after Hurricane Sandy (Knowles and Kunreuther, 2014; GAO, 2014a). As of the end of 2014, the NFIP had accrued a cumulative debt of USD 23 billion to the US Treasury, which according to the US GAO (2014a), the NFIP is unlikely to be able to repay " in the near future, if ever."

The *Biggert–Waters Flood Insurance Reform Act of 2012* has required the NFIP to build a reserve of 1% of potential losses. An assessment of 5% was charged to all policies in 2014 to build the reserve although estimates by GAO (2014b) suggest that a 25% assessment would be needed. However, the *Homeowner Flood Insurance Affordability Act of 2014* capped future premium increases at 15% making it more difficult for the NFIP to reach the reserve target in the near-term (National Academy of Sciences, 2015a). The continuation of many premium subsidies and pricing practices that exclude extraordinary events will also mean a continued structural funding deficit for the NFIP.

Figure 5.1. **NFIP premium deficit and borrowing**

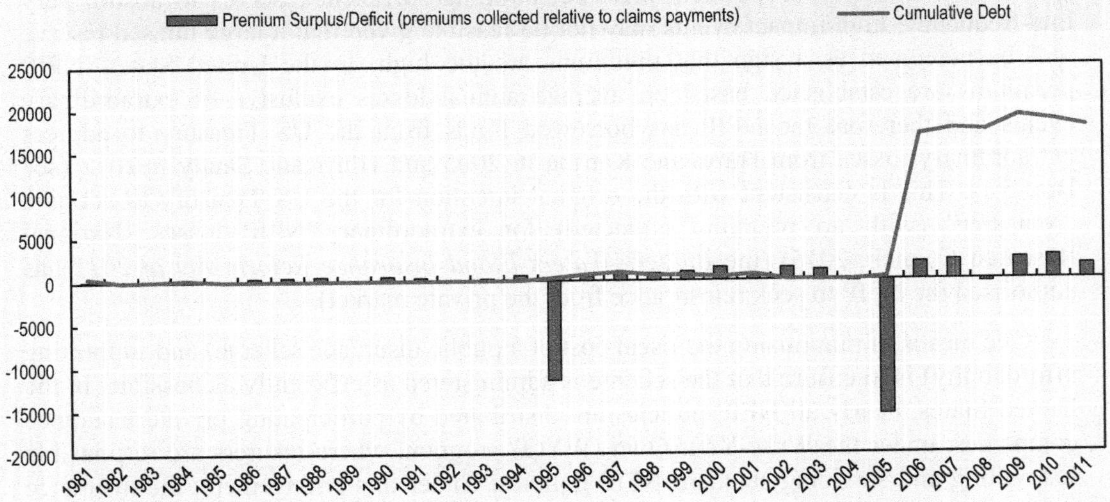


*Source*: King, 2013b

 

Reducing the number of policies and policyholders covered by the scheme could also reduce administrative expenses. Reinsurance should generally (but not always) be less costly to administer than direct insurance given the smaller number of policies and policyholders involved – although, as noted, reinsurers generally provide a ceding commission to direct insurers to account for some of the administrative burden. A community-based approach to flood insurance could also be a means to reduce administrative costs by limiting the number of policies in place. It may also be possible to generate administrative savings by simplifying the basis for claims settlement. Parametric-based triggers will be less costly to administer than indemnity-based claims settlement (although will involve basis risk to the extent that the parametric indicator used is imperfectly correlated with actual losses). In Thailand, flood insurance claims are paid based on the water level reached, rather than actual damages. In Japan, earthquake claims covered by Japan Earthquake Reinsurance are settled based on three categories of loss (total loss, half loss or partial loss) rather than through a detailed assessment of each affected structure. This allowed for claims in some areas affected by the 2011 Great East Japan Earthquake to be settled based on satellite imagery only (Tabata, 2015).

In some countries, reserves accumulated through public insurance schemes have provided a dedicated source of funding for investments in risk reduction – which should also support the management of the overall costs of public schemes over the longer-term. In France, the levy on insurance premiums is used to fund various structural and non-structural mitigation measures, including municipal risk assessments, public awareness, asset acquisition and other measures to reduce vulnerability to risks (OECD, 2014). The public National Flood Insurance Fund in the United States provides funding for mitigation investments at the household and firm-level.

Another approach to managing the potential fiscal cost of a public insurance scheme is to maximise the share of risks covered by private insurance markets (and therefore minimise the potential exposure of the public insurer). A number of countries limit the level of coverage provided by public insurance schemes for a given property/building. In the United States, NFIP coverage is limited to USD 250 000 for buildings and USD 100 000 for contents which has led to the development of a private insurance market for amounts above those limits. This can be a particularly important means for limiting the exposure of a public insurance scheme to flood risk as high-risk properties tend to be high-value properties given their proximity to coasts or other bodies of water (properties covered by the NFIP, for example, tend to be more valuable than other properties across the country (USD 160 000 median value vs. USD 220 000 to USD 400 000 in a sample of NFIP-covered properties) (King, 2013a)). Access to government insurance could also be limited for non-principal residences or based on income or some other indicator of wealth (such as an indicator of property value, such as the council tax bands used in the case for Flood Re).

A public insurer could also be limited to providing residual insurance for properties unable to access coverage in the private market. The simplest way to achieve this is by ensuring that premiums for public insurance are higher than premiums generally charged by private insurers. This is the approach taken in the United Kingdom where the pricing for Flood Re reinsurance coverage, while subsidised, has been set at levels aimed at discouraging the transfer of risk for low-risk properties. In the United States, the expected move towards risk-based pricing by the NFIP after the passage of the *Biggert–Waters Flood Insurance Reform Act of 2012* (as well as temporary relief from flood insurance rate regulation in Florida) led to the introduction of a number of private flood insurance offerings at premium rates below those mandated under the Act (Adams, 2014a and 2014b).

In the United States, a number of state wind pools have been established as residual insurers (which has the added advantage of placing extreme exposures at the level of government which has authority for many of the tools for reducing risk). Rules have been put in place to maximise private insurance companies' coverage of wind risk, including: i) ensuring that any applicants for coverage by the wind pool have sought coverage from private sector companies and have been declined; ii) requiring wind pool policies to be priced above market rates for coverage; and iii) establishing platforms that allow private insurance companies access to individual pool-backed policies up for renewal to determine whether they can provide coverage in the place of the pool. Louisiana provides financial incentives to new insurers that place 25% or more of their new business as policies taken over from state wind pool while Mississippi offers premium tax credits to insurers that write new wind and hail policies in coastal areas (Hartwig and Wilkinson, 2014). In the United Kingdom, insurance brokers have played a critical role in directing high-risk customers to insurers willing to offer coverage (according to the British Insurance Brokers Association, brokers were able to find insurance for 95% of households referred to them as a result of coverage refusals (DEFRA, 2013)). There may also be a need to limit other possible advantages of public insurance coverage over private insurance. For example, in the United States, a recent legislative proposal has been put forward to ensure that private flood insurance coverage is considered equivalent to NFIP coverage for meeting federal mortgage-related flood insurance requirements.

However, a significant disadvantage of the "residual insurer" approach is that the government will (by design) be left with the highest-risk properties (likely with premium income that is below actuarially-based rates) and without the benefit of the premium income from a large community of low-risk households. In the United Kingdom, this issue has been partially addressed through a surcharge on all domestic policyholders which provides additional income from the larger insured community. A residual insurer could also require that the private insurer retain some of the risk in order to spread the costs more broadly.

Where guarantees are provided, the government could establish an upper limit on its exposure. This could potentially be imposed on a per property basis which would improve the targeting of benefits towards homes with lower values (The Australian Government the Treasury, 2015). In Belgium, the government will pay up to EUR 280 million to each insurer if a threshold of damage is exceeded (EUR 3 million in losses plus 0.35 times premium income per insurer) (Schwarze et al., 2011). Although, as in the case of limits on compensation, the effectiveness of such ceilings depends on the difficulty in raising them (in the United States, the borrowing limit for the NFIP, which serves a similar purpose, has been raised on at least two occasions).

## 5.3 Options for risk financing and transfer

There are a number of options for financing the fiscal costs of disasters. Governments may choose to self-insure, i.e. self-finance the costs of disasters from existing budgets and any reserve funds (or through *ex post* borrowing or taxation) or transfer some of their exposure to insurance or capital markets.

The capacity of governments to self-insure against flood risk will depend on their level of exposure to flood damage as well as their ability to reallocate existing budgetary resources, increase revenue or borrow additional funds from capital markets or other lenders. A number of countries have established reserve funds or specific budgetary allocations to provide an immediately-available source of funds for disaster costs. Most countries allocate some funding for disaster costs on an annual basis (although usually only to cover a small portion of the costs). Larger (and multi-year) reserve funds may be more important for more disaster-prone countries and countries with more limited access to capital markets.

Governments with robust access to capital markets may choose to finance flood recovery and reconstruction costs through *ex post* borrowing, as part of their normal government financing operations (e.g. issuing of bonds). In these countries, raising the funds when needed is consistent with efficient balance sheet and cash management (OECD, 2015). Some countries that finance disaster costs *ex post* have made use of special taxes to support some costs. For example, in the Czech Republic, a special anti-flood tax of CZK 100 (approximately EUR 4 monthly) was imposed during 2011 on every taxpayer aimed at covering losses caused by 2010 floods (OECD, 2015). Australia imposed a one-off levy on high-income earners to address the costs of the Queensland floods (Wilby and Keenan, 2012).

Developing countries with more limited access to capital markets (or facing particularly large exposures to flood risk) can establish contingent credit lines with multilateral or bilateral development agencies. The World Bank and Inter-American Development Banks offer such facilities as does the Japan International Cooperation Agency (JICA) and a number of countries have established such lines of credit. For example, the Philippines has secured a Catastrophe Deferred Drawdown Option (Cat-DDO) contingent credit line of up to USD 500 million to provide immediate access to liquidity upon the declaration of a state of emergency (OECD, 2015) as well as a separate credit line from JICA (Laureano, 2015).

Governments may also choose to transfer some of the exposure they face for flood recovery and reconstruction costs to insurance (or capital) markets. There a variety of risk transfer mechanisms available to national and sub-national governments, including insurance for public assets, risk-pooling and catastrophe bonds (amongst others).

Insurance of individual public assets is usually more common in developing than developed countries. Some countries (e.g. Philippines, Indonesia, Australia) have dedicated publicly-owned insurance companies that provide coverage for public assets. In the Philippines, the national Government Service Insurance System (GSIS) provides coverage for public assets at the national and local levels and requirements have been established for the purchase of insurance from the resources of local reserve funds (OECD, 2015). In Indonesia, a public insurer (*PT Bangun Askrida*) is owned directly by local governments for the purpose of insuring their public assets. In Australia, a number of sub-national governments have established insurance entities to provide insurance for public assets. With the exception of roads, these insurance arrangements are considered to be providing adequate levels of financial protection (Productivity Commission, 2014). In the Czech Republic, New Zealand and Australia, the relevant government authorities are required to consider the various financial protection options for public assets, including insurance, as part of their asset management strategies (OECD, 2015). The federal government in Australia regularly reviews the insurance arrangements for public assets owned by sub-national governments, partly as a means for reducing the moral hazard created when a large share of the reconstruction costs are borne by national governments. Local governments in New Zealand are required to disclose details of their insurance arrangements in their annual reports. Alternatives to the current "light-handed" disclosure regime are being considered as part of a broader review in New Zealand on the financial assistance scheme in place for the restoration of local authority infrastructure following natural disasters.

Governments may also seek indemnity-based insurance coverage for a pool (or its full portfolio) of public sector assets. Indemnity-based insurance arrangements that cover a broad pool of assets facing diversified risks can have cost advantages over insuring the assets individually and can eliminate basis risk (a key challenge for parametric-based insurance – see Box 5.2). Mexico has established insurance arrangements to cover approximately USD 400 million in losses for the purposes of providing a source of funding for reconstruction (Ministry of Finance (Mexico), 2015). In some countries, legal and/or administrative challenges may impede the use of risk transfer by local and/or national governments. For example, in Indonesia, a 2004 law requires that a good or service be received before payment can be made, which impedes governments' ability to pay insurance premiums to secure financial protection (Jha and Stanton-Geddes, 2013).

A few national governments have entered into multi-country pooling arrangements, including a set of Caribbean and Central American countries (CCRIF), countries in Africa (Africa Risk Capacity) and a number of Pacific Island countries (Pacific Catastrophe Risk Assessment and Financing Initiative). Such pooling arrangements provide small countries with improved access to international insurance markets based on their ability to merge a set of (less) correlated risks. The CCRIF, which provides pay-outs of up to USD 100 million based on parametric triggers including storm surge and wave height related to hurricanes and excessive rainfall, has allowed for reinsurance coverage at a 54% to 59% lower cost for hurricanes relative to the cost of an individual participating country going directly to the reinsurance market for coverage (World Bank, 2012a). There may be significant potential for pooling uncorrelated country risks in other regions as well (Bayer et al, 2012).

In large countries facing diversified flood risk, pooling of risks among local governments might also offer benefits in terms of access to (re)insurance markets and costs of coverage. Based on catastrophe modelling for US hurricanes, geographic diversification of hurricane risk across US states could lead to a significant (53%) decline in the level of reserves that would be required (by insurance companies or state wind pools) to cover losses from hurricanes with return periods above 1-in-25 years (with even greater declines as the return period increases) relative to the reserves required by an individual state to cover losses (Watson, Johnson and Dumm, 2012). In New Zealand, local governments established a pooling arrangement (Local Authority Protection Programme) to help them meet their share of the funding requirements for restoring locally-owned infrastructure damaged by natural disasters (OECD, 2015). The programme's funds were depleted as a result of the 2010 Canterbury earthquake sequence although a new arrangement involving a Local Government Risk Agency is under consideration (Stobo, 2015). In the Philippines, consideration is being given to the pooling of Local Disaster Risk Reduction and Management Funds (Laureano, 2015). Legislative proposals in the United States have also recommended various approaches to pooling state disaster exposures, including an approach based on risk transfer to capital markets (Homeowners' Defense Act of 2013) and an approach based on federal reinsurance of state disaster insurance programs (Homeowners Insurance Protection Act of 2013) (King, 2013a).

A few national and sub-national governments (or related entities) have also transferred risks to international capital markets, generally through the issuance of catastrophe bonds. Catastrophe bonds are "high-yield bonds that contain a provision that may cause the principal or interest payments to be delayed or lost to investors in the event of a specified loss, such as a hurricane or earthquake" (OECD, 2011). They usually cover the higher-layers of risk (i.e. lower probability) (King, 2013a) related to truly extreme events, which helps keep the cost of such instruments more manageable. Catastrophe bonds have most often been issued by insurance companies although a few governments (Mexico) and public authorities (Metropolitan Transportation Authority of New York) have issued catastrophe bonds to transfer disaster risks (Wolfrom and Yokoi-Arai, 2015). Catastrophe bonds have also been issued by the CCRIF (with support from the World Bank) to cover its exposures. As of September 2015, Mexico had approximately USD 315 million in catastrophe bond coverage for hurricane and earthquake events within the probable maximum loss range of 98% to 99% (Ministry of Finance (Mexico), 2015). Catastrophe bond issuance for coverage of flood risk has been less prevalent than other risks (such as hurricane wind or earthquake), likely due to the complexity of modelling flood risk (see Box 5.2) and the lower levels of private insurance coverage for floods relative to other perils.

### Box 5.2. **Transfer of flood risk to capital markets**

Catastrophe bonds, the most common form of capital market risk transfer, can be designed to trigger payouts based on: i) the actual losses experienced by the bond sponsor (indemnity-based); ii) an index of industry-wide loss estimates for a given event (industry index-based); iii) modelled losses for an event of a specific magnitude (model-based); and iv) the physical characteristics of the disaster event, such as magnitude and location of an earthquake or hurricane (parametric-based). There have been relatively few issuances of catastrophe bonds covering flood risk as all of the potential types of triggers create challenges when applied to flood losses. One such bond has been issued by an insurance company for flood risk in the United Kingdom (Swiss Re, 2012). A public authority, the New York Metropolitan Transit Authority which is responsible for public transit in New York City, issued a catastrophe bond in 2013 to cover storm surge risk using a parametric trigger based on a storm surge height threshold for a named storm (Guy Carpenter, 2015).

The growth of catastrophe bonds as a risk transfer instrument faces many challenges, including high costs of structuring. In addition, the lack of insurance coverage for flood risk in many countries has limited the number of markets where catastrophe models and indices of industry-wide losses are available. Furthermore, there has been limited experience with the models and indices that are available, limiting the demand from investors for instruments based on those models and indices. From the perspective of the issuer, the complexity in understanding flood risk means that simple parametric triggers are unlikely to correlate well with actual losses, creating "basis risk" for the issuer (Swiss Re, 2012).

*Source*: Swiss Re, 2012; Guy Carpenter, 2015.

Countries (both developed and developing) can also make use of the World Bank's catastrophe bond platform to simplify the issuance of catastrophe bonds (or replicate the approach used). The World Bank's Capital-at-Risk Notes platform was used to issue a catastrophe bond to provide coverage for earthquake and tropical cyclone risk in the

Caribbean insured through the CCRIF. The World Bank issued the bond to investors and held the proceeds on its balance sheet. If triggered, the proceeds from the bond (or some portion) would be transferred directly to CCRIF. This approach allows a simplified structure for the catastrophe bond (with the World Bank (and its superior creditor status) as counterparty to both sides of the transaction) and could be accessed by other countries or replicated by governments with sufficient creditworthiness (Navarro, 2015). Where risks are pooled by government, whether as a result of the government providing direct insurance or reinsurance for disaster risks faced by households or due to a government insurer centralising insurance coverage for public assets, risk transfer (either through reinsurance or catastrophe bonds) may be an option for addressing the most extreme losses faced by the pool. Meeting the requirements of risk transfer will generally be simpler for public insurers that will usually have access to the information necessary for transferring risks to insurance or capital markets. Flood Re, for example, intends to transfer almost all of the risks it pools to international reinsurance markets (Flood Re, 2015).

## 5.4 Costs and benefits of different approaches to fiscal management of flood risk

The financial management of flood-related public exposures requires careful assessment of the relative costs and benefits of different approaches. For example, investments in risk reduction or risk transfer can reduce the future cost of recovery and reconstruction although both involve (potentially significant) upfront costs. Similarly, investment in measures to improve insurability or coverage of some flood risk by the public sector through public (re)insurance arrangements could reduce future costs related to financial assistance and compensation although consideration needs to be given to whether the reduction in future costs is worth the upfront investment.

Governments need to consider the potential exposures that they face for a range of events and then assess the most effective way to manage those exposures based on an assessment of costs and benefits. This assessment should take into account the potential for fiscal capacity (revenue, budget reallocation, taxation and borrowing capacity) to change after a severe flood event based on changing macro-economic conditions. The timing of funding requirements is also an important consideration as disasters create significant demands for short-term resources and any delay in the availability of those resources could exacerbate the economic impact of the event.

Up-front investments in risk reduction and risk transfer need to be considered against the opportunity cost of the use of those resources relative to other priorities (Productivity Commission, 2014). In general, the financial risks related to high-frequency/low-cost flooding events are most efficiently managed through risk mitigation and/or retention as transferring such risks would be costly while extreme events are generally more conducive to risk transfer (World Bank, 2012c). Risk transfer is usually an expensive option (costing multiples of the expected payout), given the need for insurance companies to be compensated for uncertainty, administration, and the maintenance of sufficient capital (World Bank, 2012c). For example, in the United Kingdom, the cost of insuring each GBP 1 of technical risk has been estimated to be GBP 1.66 (Surminski and Eldridge, 2014). These costs need to be considered against the benefits of providing liquidity in an environment where funding may otherwise be constrained.

The assessment of various approaches also needs to consider the incentives created by these approaches and the impact of those incentives on ultimate costs. For example, the broad availability of *ex post* compensation may reduce the incentives for households and sub-national governments to invest in risk reduction or risk transfer and may ultimately increase the eventual costs of financial assistance and compensation provided. High standards of physical or financial protection might also create inappropriate incentives for developing flood-prone areas. Similarly, the cost of a government reinsurance or guarantee arrangement to support risk-based insurance coverage needs to be weighed against the expected reduction in compensation needs post-event as well as the incentives that risk-based premiums could provide for risk reduction (Schwarze and Wagner, 2007). A reliance on *ex-post* financing of disaster costs may also lead to a bias against *ex ante* mitigation (Productivity Commission, 2014).

## References

Adams, M. (2014a), "Florida Flood Policy to be Offered in Six Additional States", *Carrier Management*, 1 January, *www.carriermanagement.com/news/2014/01/01/117161.htm,* accessed 6 April 2016.

Adams, M. (2014b), "Florida Private Flood Bill Now Law But Market May Take a While to Develop", *Carrier Management*, 22 June, *www.carriermanagement.com/news/2014/06/22/124899.htm*, accessed 6 April 2016.

Allianz Australia Insurance Ltd. (2011), *Natural Disaster Insurance Review Submission*, 21 July.

The Australian Government the Treasury (2015), *Northern Australia Insurance Premiums Taskforce: Interim Report*, Commonwealth of Australia, *www.treasury.gov.au/~/media/Treasury/Consultations%20and%20Reviews/Reviews%20and%20Inquiries/2015/NAIP%20Taskforce/Interim%20Report/Downloads/PDF/NAIP-interim-report.ashx.*

Bayer et al., (2012), *Insurance Instruments for Adaptation to Climate Risks: Linking Policy Agendas*, World Bank, Washington, DC

Burris, A. (2015), "Fresh legislation sparks insurer interest in Florida flood risk", *Insurance Day*, 30 June.

CBO (2006), *The Budgetary Treatment of Subsidies in the National Flood Insurance Program: Statement of Donald B. Marron, Acting Director, CBO, before the Senate Committee on Banking, Housing, and Urban Affairs 25 January*, US Congressional Budget Office, Washington, DC.

Cohn, C. and K. Holton (2015), "Britain Faces More Extreme Weather, Rising Flood Costs", *Carrier Management*, 31 December, *www.carriermanagement.com/news/2015/12/31/149422.htm,* accessed 5 April 2016.

Conrad, D. and E. Thomas (2013), *Proposal 2: Reforming Federal Support for Risky Development*, The Hamilton Project, *www.hamiltonproject.org/assets/legacy/files/downloads_and_links/THP_15WaysFedBudget_Prop2.pdf,* accessed 6 April 2016.

Cummins, J.D., M. Suher and G. Zanjani (2010), "Federal Financial Exposure to Natural Catastrophe Risk", in D. Lucas (ed.), *Measuring and Managing Federal Financial Risk*, University of Chicago Press, Chicago.

DEFRA (2013), *Securing the future availability and affordability of home insurance in areas of flood risk*, Department for Environment, Food and Rural Affairs, London.

Flood Re (2015), How it works (website), *www.floodre.co.uk/how-it-works-overview,* accessed 12 November 2015.

GAO (2015), *Flood Insurance: Status of FEMA's Implementation of the Biggert-Waters Act, as Amended (GAO-15-178)*, Government Accountability Office (February), Washington, DC

GAO (2014a), *Flood Insurance: Strategies for Increasing Private Sector Involvement (GAO-14-127)*, Government Accountability Office (January), Washington, DC.

GAO (2014b), *Overview of GAO's Past Work on the National Flood Insurance Program (GAO-14-297R)*, Government Accountability Office (January), Washington, DC.

GAO (2010), *National Flood Insurance Program: Continued Actions Needed to Address Financial and Operational Issues*, Statement of Orice Williams Brown, Director Financial Markets and Community Investment, Testimony Before the Subcommittee on Housing and Community Opportunity, Committee on Financial Services, House of Representatives, *www.gao.gov/assets/130/124468.pdf.*

GDV (2013), *Position Paper of the German Insurance Association (GDV) on the consultation on the Green Paper on the Insurance of Natural and Man-Made Disasters*, German Insurance Association (Gesamtverband der Deutschen Versicherungswirtschaft), GDV, Berlin.

Guy Carpenter (2015), "Flood Insurance Risk Study: Reinsurance Study", *National Flood Insurance Program: Report to Congress on Reinsuring NFIP Insurance Risk and Options for Privatizing the NFIP (13 August)*, Department of Homeland Security, Washington, DC.

Hartwig, R. and C. Wilkinson (2014), *Residual Market Property Plans: From Markets of Last Resort to Markets of First Choice*, Insurance Information Institute, New York.

Iceland Catastrophe Insurance (2014), *Annual Report*, Iceland Catastrophe Insurance, *www.vidlagatrygging.is/UserFiles/Documents/Annual%20Report%202014%20webpage.pdf,* accessed 6 April 2016.

Insurance Bureau of Canada (2015), *The financial management of flood risk: An international review – lessons learned from flood management programs in G8 countries*, IBC, Toronto.

Israel, N. (2013), *Inaction on Climate Change: The Cost to Taxpayers*, Ceres, Boston.

Jha, Abhas K. and Z. Stanton-Geddes (eds.) (2013), *Strong, Safe, and Resilient: A Strategic Policy Guide for Disaster Risk Management in East Asia and the Pacific*, World Bank, Washington, DC.

King, R. (2013a), *Financing Natural Catastrophe Exposure: Issues and Options for Improving Risk Transfer Markets*, Congressional Research Service, Washington, DC.

King, R. (2013b), *The National Flood Insurance Program: Status and Remaining Issues for Congress*, Congressional Research Service, Washington, DC.

Knowles, G. and H. Kunreuther (2014), "Troubled Waters: The National Flood Insurance Program in Historical Perspective", *Journal of Policy History*, Vol. 26 (3), pp. 327-353.

Kousky, C. and L. Shabman (2015), *A Proposed Design for Community Flood Insurance*, Resources for the Future, Washington, DC, December.

Kunreuther, H. and E. Michel-Kerjan (2013), "Implementing the National Flood Insurance Reform Act in a New Era of Catastrophes", *Penn Wharton Public Policy Initiative Issue Brief*, Vol. 1 (9).

Laureano, S. (2015), "Disaster Risk Financing and Insurance Strategy of the Philippines", Presented to the Global Seminar on Disaster Risk Financing: Towards the development of effective approaches to the financial management of disaster risks, 17-18 September, Kuala Lumpur, Malaysia, *www.oecd.org/daf/fin/insurance/global-seminar-disaster-risk-financing-2015.htm.*

Lindsay, B. and F. McCarthy (2015), *Stafford Act Declarations 1953-2014: Trends, Analyses, and Implications for Congress*, Washington, DC, Congressional Research Service (14 July), *www.fas.org/sgp/crs/homesec/R42702.pdf.*

Michel-Kerjan, E. (2010), "Catastrophe Economics: The National Flood Insurance Program", *Journal of Economic Perspectives*, Vol. 24 (4), pp. 165-186.

Michel-Kerjan, E. (2001), "Insurance against natural disasters: do the French have the answer", *Cahier No. 2001-007*, École Polytechnique Centre national de la recherche scientifique.

Ministry of Finance (Mexico) (2015), "Public Sector Recovery And Reconstruction Needs: The experience of Mexico's Natural Disasters Fund (FONDEN)", Presented to the Global Seminar on Disaster Risk Financing: Towards the development of effective approaches to the financial management of disaster risks, 17-18 September, Kuala Lumpur, Malaysia, *www.oecd.org/daf/fin/insurance/global-seminar-disaster-risk-financing-2015.htm.*

Moss, D. A (1999), "Courting disaster? The transformation of federal disaster policy since 1803", in K.A. Froot (ed.), *The financing of catastrophe risk*, University of Chicago Press, Chicago.

Nagamura, M. (2013), "Case study 5: Tohoku earthquake and tsunami", in Orie, M. and W. Stahel (eds.), *Insurers' contributions to disaster reduction—a series of case studies: The Geneva Reports Risk and Insurance Research No. 7*, The Geneva Association (The International Association for the Study of Insurance Economics).

National Academies of Sciences, Engineering, and Medicine (2015a), *Affordability of National Flood Insurance Program Premiums: Report 2*, The National Academies Press, Washington, DC.

National Academies of Sciences, Engineering, and Medicine (2015b), *A Community-Based Flood Insurance Option*, The National Academies Press, Washington, DC.

National Research Council (2015), *Affordability of National Flood Insurance Program Premiums: Report 1*, The National Academies Press, Washington, DC.

Nationwide (2015), "Most Small Business Owners at Risk for a Disaster", *Press release*, 31 August, *www.nationwide.com/about-us/083115-small-biz-survey.jsp*, accessed 10 December 2015.

Navarro, M. (2015), "World Bank Group Menu of Disaster Risk Financing Instruments", Presented to the Global Seminar on Disaster Risk Financing: Towards the development of effective approaches to the financial management of disaster risks,

17-18 September, Kuala Lumpur, Malaysia, *www.oecd.org/daf/fin/insurance/global-seminar-disaster-risk-financing-2015.htm.*

OECD (2015), *Disaster Risk Financing: A global survey of practices and challenges*, OECD Publishing, Paris, *http://dx.doi.org/10.1787/9789264234246-en.*

OECD (2014), *Seine Basin, Île-de-France, 2014: Resilience to Major Floods*, OECD Publishing, Paris, *http://dx.doi.org/10.1787/9789264208728-en.*

OECD (2012), *Disaster Risk Assessment and Risk Financing: A G20/OECD Methodological Framework*, *www.oecd.org/daf/fin/insurance/g20oecdframeworkfordisasterriskmanagement.htm.*

OECD (2011), "Risk Awareness, Capital Markets and Catastrophic Risks", *Policy Issues in Insurance, No. 14*, OECD Publishing, Paris, *http://dx.doi.org/10.1787/9789264046603-en*

Pollner, J. (2012), *Financial and Fiscal Instruments for Catastrophe Risk Management: Addressing Losses from Flood Hazards in Central Europe*, World Bank, Washington, DC.

Productivity Commission (2014), *Natural Disaster Funding Arrangements, Inquiry Report no. 74*, Canberra.

Regional Australia Institute (2014), *The Regional Australia Institute's response to the Productivity Commission's Draft Report on Natural Disaster Funding Arrangements*, Regional Australia Institute, Barton, Australia.

Schwarze et al. (2011), "Natural Hazard Insurance in Europe: Tailored Responses to Climate Change are Needed", *Environmental Policy and Governance*, Vol. 21, pp. 14-30.

Schwarze, R. and G. Wagner (2007), "The Political Economy of Natural Disaster Insurance: Lessons from the Failure of a Proposed Compulsory Insurance Scheme in Germany", *European Environment*, Vol. 17, pp. 403–415.

Simpson, A. (2015), "FEMA Shows Some Progress Implementing Flood Insurance Changes: GAO", *Carrier Management*, 22 February.

Standard & Poor's Ratings Services (2015), "The Heat is On: How Climate Change Can Impact Sovereign Ratings", *Ratings Direct*, 25 November.

Stelmakowich, A. (2013), "Overland Overhaul", *Canadian Underwriter*, August, *www.canadianunderwriter.ca/news/overland-overhaul/1002551174/?type=Print%20Archives*, accessed 11 January 2016.

Stobo, C. (2015), "Risk Sharing Among Local Governments – New Zealand's Experience", Presented to the Global Seminar on Disaster Risk Financing: Towards the development of effective approaches to the financial management of disaster risks, 17-18 September, Kuala Lumpur, Malaysia, *www.oecd.org/daf/fin/insurance/global-seminar-disaster-risk-financing-2015.htm.*

Surminski, S. and J. Eldridge (2014), "Flood insurance in England – an assessment of the current and newly proposed insurance scheme in the context of rising flood risk", *Centre for Climate Change Economics and Policy Working Paper No. 161/Grantham Research Institute on Climate Change and the Environment Working Paper No. 144*, Centre for Climate Change Economics and Policy, Leeds.

Swiss Re (2015), *The natural catastrophe protection gap in Italy: time for action*, Swiss Reinsurance Company Ltd., Zurich.

Swiss Re (2012), *Flood – an underestimated risk: Inspect, Inform, Insure*, Swiss Reinsurance Company Ltd., Zurich.

Tabata, H. (2015), "Earthquake Insurance in Japan", Presented to the Global Seminar on Disaster Risk Financing: Towards the development of effective approaches to the financial management of disaster risks, 17-18 September, Kuala Lumpur, Malaysia, *www.oecd.org/daf/fin/insurance/global-seminar-disaster-risk-financing-2015.htm.*

Tournier, B. (2011), *Municipal Market Comments Natural Disasters Not a Threat to Muni Ratings*, Montage Investments, 25 August.

Watson, C., M. Johnson and R. Dumm (2012), "The Impact of Geographic Diversity on the Viability of Hurricane Catastrophe Insurance: Final Report", a project completed for the Florida Catastrophic Storm Risk Management Center, Florida State University, November 2012,
*www.stormrisk.org/sites/default/files/GeoDiversity_Final_Report.pdf.*

Weiss, D. and J. Weidman (2013), *Disastrous Spending: Federal Disaster-Relief Expenditures Rise amid More Extreme Weather*, Center for American Progress (29 April), *www.americanprogress.org/issues/green/report/2013/04/29/61633/disastrous-spending-federal-disaster-relief-expenditures-rise-amid-more-extreme-weather/,* accessed 5 April 2016.

Wilby, R. and R. Keenan (2012), "Adapting to flood risk under climate change", *Progress in Physical Geography*, Vol. 36 (3), pp. 348–378.

Wolfrom, L. and M. Yokoi-Arai (2015), "Financial instruments for managing disaster risks related to climate change", *OECD Journal: Financial Market Trends. http://dx.doi.org/10.1787/fmt-2015-5jrqdkpxk5d5.*

World Bank and Thai Ministry of Finance (2012), *Rapid Assessment for Resilient Recovery and Reconstruction Planning*, World Bank, Bangkok.

World Bank (2012a), *Advancing Disaster Risk Financing and Insurance in ASEAN Member States: Framework and Options for Implementation: Volumes 1 and 2*, World Bank, Washington, DC.

World Bank (2012b), "Caribbean Catastrophe Risk Insurance Facility: Implementation Completion & Results Report", World Bank, Washington, DC, 12 July.

World Bank (2012c), *Disaster Risk Financing and Insurance in Sub-Saharan Africa: Review and Options for Consideration*, World Bank, Washington, DC.

# *Chapter 6*

## Designing a disaster risk financing strategy for flood risk

*This chapter provides a summary of the main recommendations in the report for the purposes of designing a disaster risk financing strategy for the financial management of flood risk.*

The effective financial management of flood risk is a complex public policy challenge for countries faced with significant exposures to flood risk and/or limited capacity to manage the financial impacts of that risk, particularly in the context of a changing climate. Managing flood risk requires careful consideration of the costs and benefits of different approaches, including managing the incentives created by different forms of intervention.

The OECD is preparing a draft Recommendation on the development of disaster risk financing strategies[1] which provides some overarching design principles and outlines the main components of an effective strategy, including:

- promoting comprehensive risk assessment processes that allow for the estimation of exposures and the identification of financial vulnerabilities;
- supporting the management of the financial impacts of disasters by all segments of the population and economy and encouraging the development of risk transfer markets; and
- effectively managing the financial impacts of disasters on public finances.

This chapter will outline how these can be applied in developing a strategy for the financial management of flood risk, based on the findings of this report.

## 6.1 Estimating exposures and identifying financial vulnerabilities

The accurate assessment of flood risk is an essential prerequisite for the effective financial management of flood risk. A comprehensive understanding of exposures to flood risk for different locations is necessary for effective land-use planning, the development of building and design standards to protect against inundation, and for assessing the relative costs and benefits of investments in risk reduction measures. It is also critical for reducing the impact of floods when they occur by providing emergency managers with the information they need to intervene (e.g. the placement of temporary emergency dams or for evacuating communities). It is also a prerequisite for the transfer of flood risk to (re)insurance and capital markets.

However, as noted in Chapter 3, the assessment of flood risk is complicated by a number of factors, including the broad range of causes of floods, the significant differential in impacts based on small changes in water level, and the uncertainty related to the nature of flood risk in a changing climate. Climate change is expected to have important implications for the nature of flood risk going forward as a result of changes to the frequency of heavy precipitation events, the range and frequency of cyclones, and the rise in sea-levels which needs to be accounted for in assessing future flood risk. A number of countries are adding climate change allowances in assessments of flood risk. For example, in Australia, the Queensland Inland Flooding Study recommends a 5% increase in rainfall intensity for each 1°C increase in global warming while in New South Wales, a 10-35% increase in extreme rainfall is recommended in sensitivity analyses of future projections (Wilby and Keenan, 2012).

Flood hazard and risk maps are becoming more broadly available (almost all surveyed countries indicated that such maps are in place and many are updated on a regular basis (usually 6 years)). However, a number of countries still face challenges in terms of the quality and consistency of flood maps which makes it difficult to construct a consistent and accurate representation of flood risk for many countries. Probabilistic flood models that provide the estimates of damage and losses necessary for underwriting insurance coverage and for accurately assessing costs and benefits of different risk reduction measures are only available for a few countries - driven largely by the demand for such modelling capability from private insurance companies.

Where probabilistic flood models are not available, experience from past events can provide an (imperfect) source of information for understanding flood exposure. Insurance companies can be an important source of information on past impacts from flooding which can be used to support land-use planning and decisions on risk reduction. For example, in Norway, the private insurance sector has made data on past losses at the level of individual structures available to municipalities as a means of supporting their understanding of flood exposure (Ebeltoft and Nussbaum, 2016). Increasing access to satellite technology also provides an opportunity to assess the impacts of past events. A commitment to undertaking post-disaster loss assessments for significant events, as occurs in some countries, is another means of improving the availability of the data necessary for accurate risk assessment.

## 6.2 Supporting the effective financial management of flood risk

Countries with broad insurance coverage for disaster risks tend to face more limited economic disruption as a result of disaster events. Insurance provides a timely source of financing for reconstruction (in many cases, sourced from international reinsurance markets) and reduces the potential costs to the public sector in covering uninsured losses. However, as outlined in Chapter 3, there are a number of challenges to the insurability of flood risk, including the size of expected losses, uncertainty in the quantification of exposures and limited ability to establish a pool of diversified risks – all of which can lead to high prices for insurance coverage. At the same time, the willingness-to-pay for flood insurance is limited by low levels of risk awareness, misunderstandings about coverage and expectations of government assistance which creates a market failure that reduces the level of financial protection against flood risk and leads to a significant financial protection gap.

Governments have a critical role in supporting the insurability of flood risk through effective land-use planning, (including both restrictions on development and allowances for natural flood protection mechanisms that can enhance water absorption and protect against storm surge) and by investing in - and providing financial support for - structural (community-level) and household risk reduction measures which can be highly effective in reducing flood risk. In decentralised countries, national governments have an important role in ensuring that local governments have the right incentives (and authorities) to take flood risk into account in local planning and investment decisions. A few countries take the availability and/or affordability of flood insurance coverage directly into account when making decisions on where to target investments in risk reduction. In some countries, explicit commitments from government to implement strict land-use controls and finance risk reduction investments have been sought by the private insurance sector as a condition for offering flood insurance on a broad-basis. High-risk areas, often developed before the true level of flood risk was known, should be a particular focus for risk reduction given the difficulty of providing a viable insurance offering to households in those areas.

The form of insurance coverage can have important implications for the level of take-up. The automatic extension of general property insurance coverage to include protection against flood damage as well as approaches that include flood coverage as the default option for insurance policies have led to significantly higher levels of flood insurance penetration. Requirements for flood coverage as a condition for mortgage financing have also been successful in encouraging take-up (when effectively enforced). Where insurance coverage for flood is an optional add-on to property policies, investments in improving public understanding of flood risk and the need for financial protection will likely be necessary for generating sufficient demand for flood insurance.

There is some evidence that forms of risk communication that focus on return probabilities within shorter time periods, build on recent flood experience and provide estimates of the potential level of flood damage may be more effective in encouraging households and businesses to seek financial protection. Minimising misunderstandings about the scope of flood coverage as well as clarifying the extent of possible public disaster assistance may also be important to increasing the demand for flood coverage.

In a number of countries, flood insurance coverage is provided by the public sector or through a public-private partnership, whether as a result of limited private insurance sector appetite for flood exposure or an explicit decision by government to intervene in order to achieve other policy objectives (e.g. broad availability and affordability of coverage or solidarity in terms of loss-sharing across regions). If administered efficiently, such schemes can support affordability by reducing the cost of providing coverage.

Whether private or public, automatic or optional, the contribution of insurance to the financial management of flood risk will be enhanced where insurance contributes to risk awareness and encourages risk reduction. Risk-based premiums, including the availability of premium discounts for risk reduction measures, can provide an important price signal on the level of exposure and a financial incentive for risk reduction. Premiums that do not reflect risk, including as a result of premium subsidies, risk encouraging development in flood-prone areas and increasing the overall level of flood exposure. The regulatory framework for insurance companies, including the framework for competition/market entry, premium pricing, reinsurance arrangements and/or asset allocation, can all have an impact on the capacity of the insurance sector to provide flood insurance coverage, meet obligations to policyholders, and support risk reduction through their investment decisions.

Public financial assistance for sub-national governments and households (and potentially businesses) affected by flood events could be essential for reducing hardship and minimising economic and social disruption. The rationale for such assistance may be particularly strong for vulnerable households living in high-risk areas where the level of flood risk may not have been known when the area was developed. However, extensive and/or poorly defined financial assistance can lead to moral hazard and reduce the incentives for sub-national governments and households, who often have the greatest ability to reducing the potential damage and loss from flood events, to invest in risk reduction and secure financial protection. Despite the existence of hazard maps, less than 40% of the respondents to the OECD survey provided an estimate of the share of the population facing flood risk, suggesting that more could be done in terms of identifying vulnerable segments of the population. Higher levels of insurance coverage among households, businesses and sub-national governments can make an important contribution to reducing the need for public financial assistance and therefore reduce the potential burden on public resources. A number of countries have aimed to address moral hazard by tying the receipt of public disaster assistance to the purchase of insurance.

## 6.3 Managing government exposures

The public sector is exposed to flood risk through the costs of relief and recovery, reconstruction of public assets, payments as compensation and financial assistance to individuals, business and/or sub-national levels of government as well as any costs related to public (re)insurance schemes that provide coverage for flood damages and losses. There are a variety of ways to manage these public sector exposures, including through cost-effective investments in risk reduction, efforts to minimise the cost of financial assistance and/or public insurance schemes and by securing financial protection for some part of the overall exposure. Careful management of the scope of financial assistance and public insurance arrangements as well as related operational costs can make an important contribution to minimising the overall cost of such arrangements.

In general, the most significant costs relate to the rebuilding of public infrastructure, often financed through cost-sharing arrangements between national and sub-national governments (that are often responsible for a large share of public infrastructure assets). Financial assistance to sub-national governments, taking into account the relative fiscal capacity of each level of government, can be critical for supporting the ability of sub-national governments to manage the financial impacts of flooding. However, national governments need to ensure that such assistance does not discourage investment in risk reduction or financial protection at the sub-national level. National governments could vary cost-sharing arrangements based on the level of adherence to robust land-use restrictions and building codes or could organise compensation programmes as insurance arrangements with premiums that vary based on risk, rates of co-insurance and/or coverage levels.

Where governments provide insurance coverage for flood risk, whether as a direct insurer, reinsurer or guarantor, public exposure to flood risk can be minimised by maximising the share of risk transferred to the private sector. This can be achieved by limiting the availability of public insurance to residual markets (where private insurance is not available or affordable), limiting the amount of coverage provided through a public insurance arrangement and/or requiring private insurers to retain a share of any risks transferred to a public reinsurer. However, a residual insurance arrangement will result in the government taking-on the worst risks (without the benefit of the premium income from good risks - although this can be partially mitigated by imposing cross-subsidy on the good risks [e.g. through a surcharge]).

Governments with access to international capital markets may have limited incentives to transfer fiscal risk to insurance markets although the use of insurance may still be beneficial as a means for encouraging risk reduction. The use of other risk transfer mechanisms by governments to manage the financial impacts of flood risk is limited. The transfer of flood risk to capital markets has been particularly challenging due to the complexity of flood modelling and the more limited capital market acceptance of instruments based on flood model losses. The pooling of flood risk across and within countries may offer opportunities for improving access to – and the affordability of – reinsurance coverage for public sector exposures to flood.

### Box 6.1. **Key policy messages for the design of a disaster risk financing strategy for flood risk**

- **The ability to quantify exposure to flood risk is a prerequisite to the effective financial management of flood risk and a necessary input for assessing the costs and benefits of different approaches to risk reduction and for transferring risk to (re)insurance capital markets.**
  - Assessments of flood risk need to account for the uncertainty related to the impacts of climate change on flood exposure.
  - The insurance sector is an important source of information on exposure and past losses that should be leveraged by governments for risk assessment purposes. The development of private flood insurance markets has also been a key driver for the development of flood modelling capacity.
- **Government involvement is key in supporting the insurability of flood risk.**
  - Minimising exposure to flood risk through effective land-use planning and investments (or encouraging investments) in risk reduction at the community and household level are critical for improving the insurability of flood risk. In decentralised countries, national governments need to ensure that local governments have the right incentives and authorities to take flood risk into account in local planning and investment decisions. Challenges in terms of the availability and affordability of insurance coverage in a given area are an important indicator of where risk reduction investments should be focused.
  - Insurance arrangements that make it more difficult for policyholders to exclude flood coverage in their general property insurance policies have been more successful in achieving higher levels of flood insurance penetration. Where coverage for flood is optional, investments in raising public awareness of flood risk and the need for financial protection will likely be necessary. Insurance companies, associations and brokers have a clear role to play in raising awareness among their customers.
  - Whatever the form of insurance coverage, the contribution of insurance to the financial management of flood risk will be maximised where insurance promotes risk reduction. Risk-based premiums and premium discounts for risk reduction measures can make an important contribution to maximising the benefits of insurance.
  - The regulatory framework should be designed to support the contribution of insurance to the financial management of flood risk by not establishing any (unwarranted) restrictions in areas such as asset allocation, risk transfer and premium-setting. Policies to support the development of viable insurance markets, taking into account different country circumstances, can make an important contribution to reducing the financial protection gap. International organisations should support this objective in their country programmes.
- **Effective coordination across government is critical for establishing an integrated approach to the financial management of flood risk that considers the best-use of limited public resources and takes into account the costs and benefits of different approaches (including the incentives created by different interventions).**
  - The exposure of the public sector to flood costs can be minimised by carefully managing the scope of financial assistance and public insurance arrangements and by maximising the share of risk transferred to the private sector.
  - Given the range of policy tools that need to be considered, a holistic approach to the financial management of flood risk requires effective coordination across government, including across levels of government, supported by strong leadership aimed at addressing the financial vulnerabilities created by exposure to flood risk.

Ultimately, the effective financial management of flood risk requires governments to consider the best-use of their limited resources, taking into account the cost and benefits of different approaches including the incentives created by different interventions. In particular, governments need to examine the causes of under-investment in risk reduction prevalent in most countries and the best means to correct this imbalance. Achieving this will require effective coordination across government departments and different levels of government along with strong leadership aimed at addressing the financial vulnerabilities created by flood risk.

## Note

1. As the result of a review of the *Recommendation of the OECD Council on Good Practices for Mitigating and Financing Catastrophic Risks (2010)*, the OECD is developing a Recommendation on Disaster Risk Financing Strategies to replace the original. The draft text for the new Recommendation was made available for public comment until 15 April 2016 (see: *www.oecd.org/pensions/public-consultation-drf.htm*). At the time of writing, a draft Recommendation is being prepared for adoption by the OECD Council.

## References

Ebeltoft, M. and R. Nussbaum (2016), "The role of Private Sector: The National Insurance Association - Comparative analysis between France and Norway", Presented to the 7th Technical Workshop on EU Loss Data, 10-11 March, Ispra, Italy, *http://drmkc.jrc.ec.europa.eu/DesktopModules/CritechDocumentItems/API/CritechDocumentItems/Download?contentItemId=544&forceDownload=false,* accessed 12 April 2016.

Wilby, R. and R. Keenan (2012), "Adapting to flood risk under climate change", *Progress in Physical Geography*, Vol. 36 (3), pp. 348–378, Sage.